Marconi
on the
Isle of Wight

Marconi
on the
Isle of Wight

Tim Wander

First published in Great Britain 2013.

Parts of this book were first published in paperback, as a limited (500 copies) edition *'Marconi on the Isle of Wight'* by TRW Design and Print.   ISBN 978-0- 9538671-1-0
© T.R. Wander 2000

ISBN 978-0-7552-0720-6

Authors OnLine Ltd.
19 The Cinques,
Gamlingay, Sandy,
Bedfordshire, SG19 3NU.
England.

A CIP catalogue record for this book is available from the British Library.

The Author asserts the moral right to be identified as the author of this work.

The author can be contacted via:

www.2mtwrittle.com

This book is also available in e-book format, details of which are available at:
www.authorsonline.co.uk.

For Patch.....

Still running up mountains.....

and for Ellie
who will always be up there...

Guglielmo Marconi, c. 1896

'The air was full of the promises of miracles'

Marconi on the Isle of Wight

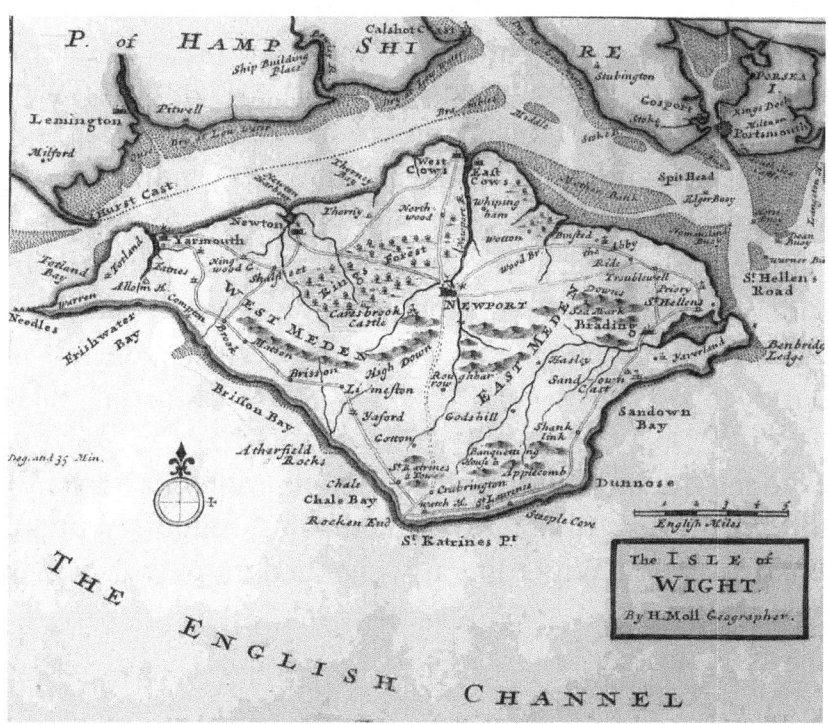

The Isle of Wight is England's largest island, located between two and five miles off the southern coast of England, separated from the mainland by a strait of water called the Solent. The Island has a rich history, including a brief status as an independent kingdom in the 15th century. The Island has some exceptional wildlife and is one of the richest locations of dinosaur fossils in Europe.

Today the Island has many resorts which have been popular holiday destinations since Victorian times. Queen Victoria built her much-loved summer residence and final home, Osborne House at East Cowes. As a result the Island soon became a major holiday resort for fashionable Victorians including Alfred, Lord Tennyson, photographer Julia Margaret Cameron, poets Swinburne and Keats, author and playwright J.B. Priestley and Charles Dickens (who wrote much of David Copperfield there) along with many members of European royalty. Charles Darwin stayed at the Kings Head Hotel in Sandown during the summer of 1867 and it is believed that this is where he began his 'Origin of Species' assessment. The island's maritime and industrial history encompasses boat building, sail making, the manufacture of flying boats, the world's first hovercraft and the testing and development of Britain's missile systems. The *AA Ordnance Survey Leisure Guide* described it:

> 'High chalk cliffs, sandy bays, bird haunted mud flats and windswept coast - the Isle of Wight, never remote but always apart, has drawn those who seek what Karl Marx called 'a little paradise'. It was a place of inspiration for Tennyson, of quiet retreat for Queen Victoria and of imprisonment for King Charles I. Today's 'overners' from the mainland will find pretty villages, seaside resorts, manor houses, Victorian churches, lighthouses and even vineyards, all set in a varied landscape.'

In 1898 the young Italian inventor, Guglielmo Marconi came to the Isle of Wight to build the world's first permanent wireless station. Marconi's time and experiments on the Isle of Wight literally changed the world, giving birth to the modern wireless age.

Contents

Photographs and Illustrations

Front Cover: The Royal Needles Hotel, Alum Bay Wireless Station, 1897
Front Cover Right: The Royal Yacht *Osborne*
Front Cover Left: Guglielmo Marconi and his wireless equipment, c.1899

Back Cover: The Niton Wireless Room, 1900
Back Cover Right: George Kemp, Marconi's lifelong assistant
Back Cover Left: The Niton Wireless Station, c.1899

Inside Back Cover: The author photographed in front of Marconi's Isle of Wight Niton station,
                   Seated on the aerial base, May 2011

Guglielmo Marconi, c. 1899
Isle of Wight Map, c. 1900

Chapter 1. Prelude to Wireless

1. Villa Griffone c. 1894 *(MWT)*
2. Annie Marconi with the young Alfonso and his brother, Guglielmo
3. Annie and Guglielmo Marconi
4. Marconi's early equipment in Italy *(MWT)*
5. Marconi's laboratory, Villa Griffone, Italy 1895 *(MWT)*
6. Marconi with his wireless telegraphy equipment *(MWT)*
7. William Preece
8. Henry Jameson-Davis *(MWT)*
9. Marconi's trials on Salisbury Plain
10. George Kemp *(MWT)*
11. Righi Oscillator Transmitter
12. Marconi's parabolic transmitter, Salisbury Plain
13. Marconi's parabolic receiver, Salisbury Plain

Chapter 2. Alum Bay - The World's First Permanent Wireless Station

Chapter 3. Bournemouth - The Madeira House Hotel

Chapter 4. By Royal Command

Chapter 5. Research and the Haven Hotel

Chapter 6. Niton, Tuning and a Trans-Atlantic dream

Appendice Photographs

Appendix 1. The Isle of Wight and South Coast Sites Today

68. The new Royal Needles Hotel
69. Royal Needles Hotel, burnt out shell. c. 1915
70. Royal Needles Hotel, 1928
71. Royal Needles Hotel, c. 1908
72. Alum Bay, c. 1924
73. Alum Bay monument
74. Alum Bay monument plaque
75. Alum Bay site of Royal Needles Hotel, 2012
76. Alum Bay site of Royal Needles Hotel, 2012
77. Alum Bay aerial view, c. 2000
78. Alum Bay aerial view, c. 2000
79. Alum Bay chairlift
80. Alum Bay, Marconi's Bar, 2012 showing original hotel stable block
81. Alum Bay, Marconi's Bar, 2012
82. Alum Bay, Marconi's Bar, Interior 2012

83. Totland Post Office memorial board and text, 2012
84. Totland Post Office, 2012
85. Inside Totland Post office, 2012

86. Fort Victoria, 2012
87. Fort Victoria Pier, 2012

88. Osborne House Golf Club
89. Site of Ladywood Cottage, 2012

90. Niton Lighthouse and Knowles Farm
91. Knowles Farm
92. Knowles Farm
93. National Trust Plaque, Niton
94. Niton wireless station plaque
95. Niton wireless station, from seaward side, with aerial base 2011

Glossary

Photo Credits

All photographs marked *(MWT)* are reproduced by kind permission of what I knew as **GEC-Marconi Ltd**. In this text I have, for the sake of brevity and clarity usually referred to the **'Marconi Company'** which was founded by Guglielmo Marconi in 1897 as the **Wireless Telegraph & Signal Company Ltd**. It was renamed **Marconi's Wireless Telegraph Company Ltd** in 1900 and **The Marconi Company (Ltd.)** in 1963.

Other photographs © Authors Collection, except number 77 & 78 - © Alum Bay Needles Park and 59, 60, 61, 104 and 104 used by kind permission of Barry and Kathleen Roberts.

Every effort has been made to fulfil requirements for reproducing copyright material although most of the images are now at least 115 years old, some much older. Some images (or other media files) are now in the public domain because their copyright has expired. This applies to Australia, the European Union and those countries with a copyright term of life of the author plus 70 years. The author would be glad to rectify any omissions at the earliest opportunity.

Authors Note - Wireless, Radio and Hertzian Waves

In this text the terms 'wireless' and 'radio' mean exactly the same thing. Surprisingly it was the word 'radio' that first came into use even before Heinrich Hertz proved the existence of electromagnetic waves in 1888, but despite this very early usage 'radio' has always been regarded as the modern version of the term 'wireless.' Hence in this story, the terms 'wireless waves', 'radio waves', 'etheric waves', 'Marconi waves', 'Hertzian waves' and 'electromagnetic waves' all mean essentially the same thing.

This story of Marconi on the Isle of Wight starts over one hundred and twenty years ago, so for the most part I use the term wireless, although by 1920 radio had become the accepted term. I was recently reminded that a young Italian called Guglielmo Marconi once called his Company..... Marconi's *Wireless* Telegraph Company Ltd, and how could he be wrong?

Acknowledgements

A small part of this book and some of the photographs were first produced as a Limited Edition brochure of just 500 copies produced for the centenary celebrations of the closure of the world's first permanent wireless station held at Alum Bay in May 2000. This was a fun day out, the local amateur radio club contacted HMS *Ark Royal,* the QE2 and even Concorde on final approach to New York airport. There was a marching band, helicopters, and even an illustrated lecture called *Marconi on the Isle of Wight.*

The books that accompanied that talk have now become something of a collector's item, and I still receive requests for copies over twelve years later. A few years ago I found a few of the slim volumes and they were snapped up quickly. Occasionally they still appear for silly sums on internet auction sites and I suppose that this may in fact be a form of complement, or at least a measure of the interest that still surrounds the name Marconi.

So twelve years on I decided to take the original text, rework and re-research it all and tell the complete story of what happened on the Isle of Wight and just across the Solent on the South coast of England. This is actually part of a much larger project that had been ongoing for at least the past fifteen years to fully document the first five years of Marconi's career. Hence to just tell the story of *Marconi on the Isle of Wight*, the Bristol Channel trials, Ireland and American yacht demonstrations, South Foreland, English Channel, Boer War and Poldhu Point are only briefly mentioned to carry the Island story on. One day this huge book, or books, called ***A Kind of Magic*** will see the light of day. If you enjoy this story, perhaps you will consider reading it as well.

Back then my thanks went to Michael Paskins for his hard work in proof reading the first text and of course, through it all and all the books, to my wife Judith for, as always, typing above and beyond the call of duty. From that first small 'edition' my thanks also went to what was then the Marconi Company Ltd and its then archivist Roy Rodwell for all his help.

Also my thanks to all the people who have met me and corresponded with me throughout the world, often by instant electronic messages that Marconi could only have dreamed of. Thank you all for your interest, help and advice during my long and ongoing search for the wireless heritage of this country.

Tim Wander,
November 2012

PREFACE

In recent times we have lived through and experienced the society transforming personal computer, internet, mobile telephone and now social networking revolutions. These technological explosions have affected every area of society and touched every aspect of our lives. These new technologies have fundamentally changed the way we receive news and information, communicate with family and friends, and they have radically altered the way we do business.

But just over one hundred years ago the advent of wireless had a far bigger social and especially emotional impact.

At the dawn of the 20th Century and for many more years some people even denied the very existence of wireless waves, while others believed wireless signals to be of satanic origin, or that it was a kind of magic, black or otherwise. But this was just one small facet of the all-pervading magnetism and magic of wireless.

For the past thirty years I have been fascinated with the development of wireless communication and the birth of radio broadcasting, both of which were part of what I have called Britain's second industrial revolution.

The story of wireless is really the story on one man called Guglielmo Marconi. It is a fascinating study of how a foreigner with little or no formal education could fly in the face of established scientific opinion, stand toe to toe with the greatest minds of late Victorian England and prove them all wrong.

Working against many pressures, financial, personal and even political, the young Marconi faced scorn and sometimes even derision from the established scientific community. He fought the vagaries of the British weather, industrial espionage, fraud, blackmail, theft, Post Office and government interference and showed that it was possible to build a reliable system of communication using wireless waves.

The most amazing thing of all was that Marconi built the world's first permanent wireless station on the Isle of Wight and later, at Niton, he made the breakthroughs in tuning technology that would change the world forever.

This then is the story of Marconi on the Isle of Wight and the amazing things that happened there.

Tim Wander,
November 2012

INTRODUCTION

A Need to Communicate

Since the earliest days of civilisation the fastest means of communicating a message was some form of visual signal, but only if the weather was clear. Otherwise the message transit time was limited by the speed of a sailing ship or the stamina of a galloping horse.

The development of cable telegraphy (from Greek *tele*, 'far' and *graphein*, 'writing') in the 1830s and 1840s changed the world. Messages could now be sent and received almost instantly by the dot and dash tones of Morse code across vast distances and eventually even oceans, when in 1850 the first submarine telegraph line was laid between England and France. Over the next decades the world was linked for the first time by transoceanic submarine cables.

The Isle of Wight started its long association with communication systems when the first undersea telegraph cable was installed on 14th October 1853 connecting the Isle of Wight to the mainland by an early telegraph cable and its associated land lines. The lines ran from Southampton to Osborne House, the Queen's residence on the Island.

In 1876 the telephone was invented, eventually replacing electric Morse code tones and the need for highly skilled operators with the simple human voice. The new telephone hit the papers in January 1878 when Alexander Graham Bell demonstrated his invention to a curious Queen Victoria on 14th January at Osborne House, East Cowes in the Council Room with calls placed to London, Cowes and Southampton. These were the first long distance calls in the UK. The Queen simply stated that: 'It is rather faint and one must hold the tube rather close to one's ear.'

In March 1882, William Preece, the Chief Engineer of the British Post Office, installed an *induction* based but none the less *wireless* communication system across the Solent, between the UK mainland and the Isle of Wight, when the submarine telegraph cable failed at Hurst Castle. Preece was to try and develop his system for the next twenty years but was overtaken by Marconi's system that used radiated Hertzian waves.

As the new century approached Victorian England toasted the advent of a new scientific age. Between 1885 and 1889 the German physicist Heinrich Rudolf Hertz had demonstrated that the Scottish physicist and mathematician James Clerk Maxwell's theoretical electromagnetic waves, or as we now call them radio waves were a fact. He could generate them, receive them and even measure their properties and they quickly became known as 'Hertzian Waves' or even 'wireless waves.' Hertz was one of the best scientists of his generation, but he died in 1892 at the very young age of 36 and never saw any practical use for his laboratory experiment.

Heinrich's 'Hertzian waves' were still little more than a scientific curiosity when a number of other inventors and scientists, including Professor Oliver Lodge, started organising lectures and practical demonstrations repeating his work. Lodge sent his first short range wireless transmissions in 1894 in Oxford, but like many others, he also failed to see any commercial potential in his experimental results.

Part of the problem was that the prominent scientists of the time had announced the curvature of the earth would severely limit the range of these new waves and consequently they had no practical purpose. This impracticality was increased as it was believed that the powerful waves from any high powered long distance station would swamp the feebler waves from ordinary ship and shore stations, producing nothing but chaos in any receiver. The established scientific community were adamant that wireless communication by electromagnetic waves had no future and could never challenge the well-developed cable telegraph system and network.

In 1896 a young Italian inventor arrived on Britain's shores with a dream of transmitting messages in all weathers and at any time and place, without the use of wires. He set out on a determined, almost obsessive path to prove that his laboratory experiments were now the basis of a reliable, long distance communication system.

Yet from the moment he first set foot in England, Marconi's public demonstrations and private tests were each designed to move his system one step further, both in technological development and in commercial understanding and acceptance.

When he arrived on these shores in February 1896 he had just two boxes filled with some barely working and rather crude wireless equipment, fresh from his experimental work bench that had already been damaged by wary customs officers.

Marconi had an introduction to a distant cousin that he had not seen since he was a child, but he no staff, no finance, no offices or workshop, no scientific training, formal education and perhaps most importantly for Victorian England, no reputation. What he did have was an enormous passion and an almost unbelievable belief in himself, his ideas and his system.

At times it was only the charisma of a young Italian inventor that kept interest alive. Guglielmo Marconi was convinced that the eminent men of the British scientific establishment were wrong and that reliable long distance communication was just a matter of increasing transmitted powers, building bigger aerials and improving the sensitivity of the receivers. This belief was soon backed by his repeated observations of successful transmissions to ships lying below the horizon, suggesting that the propagation of wireless waves was far more complex than the 'straight line' or 'line of sight' theorists believed.

The problem for Marconi was that the great Heinrich Hertz had shown that his invisible waves obeyed exactly the same laws of reflection and refraction as did light waves, and that in fact the only fundamental difference between the two was one of frequency. These conclusions had also been verified time and again by Marconi and other scientific workers. Yet equally beyond dispute was the fact that the data being amassed from Marconi's demonstrations and experiments at numerous wireless stations around the world showed a steady upward progress in the ranges that could be reliably achieved. For some reason Hertzian waves were breaking the 'straight line' rule. At first the difference between theory and observation was small, but it soon rose to double the predicted figure. When distances of eighty miles could be guaranteed, and Marconi could reliably communicate with ships and other stations far below the horizon this anomaly could no longer be ignored. Still the common scientific opinion of the day simply chose to doubt the authenticity of these experimental results.

For the first five years after arriving on England's shores, Guglielmo Marconi had two driving ambitions that fuelled his research.

He was determined to quash the persistent rumours regarding the accuracy of the ranges he claimed for his system. A practical and reliable range of at least 25 miles in all weathers was essential to his dream of providing systems to safeguard ships at sea. By 1897 his system could do that and more every day and in every type of weather. It is easy to forget that before Marconi and the widespread introduction of his reliable wireless communication system, any ship, large or small, commercial, civilian or military effectively disappeared as soon as it left sight of land. If the ship did not arrive, its crew, passengers and cargo simply vanished for all time without trace.

Marconi's second goal was perhaps more ambitious. He was determined to set his system in direct competition with the long distance undersea telegraph cable companies. By 1899 wireless communication was still largely retracing the path blazed by the cable telegraph half a century earlier, developing point to point communication, albeit now without the need for expensive connecting wires.

But developments moved rapidly. As the new century approached there were the beginnings of talk about innovations which moved beyond what the cable telegraph could do. People started to understand that the new 'wireless' could indeed be useful in aiding safety at sea. Marconi also saw a role for the new system in 'the warfare of the future', and also the potential to someday compete with the telephone in providing personal communication.

By 1898 Marconi had already built the world's first permanent wireless station and new stations were being established on lightships and on the English east and south coasts. Marconi's wireless system instigated the first ship to shore message, the first shipwreck rescue, first use of the international distress signal and undertook the first transmissions to France across the English Channel. He established the first use of wireless in the air and after extensive trials with the Royal Navy he sent his fragile new system to face the harsh environment and demands of the military during the Boer war in South Africa for the first time.

Throughout this period, Marconi was determined to maintain the technological integrity of his Company, never publicising new inventions until they had been fully proved or making promises which he could not substantiate. Through it all the young Italian displayed a considerable flair for publicity and a deft hand with the press, public, military authorities and even the doubting scientific community.

Guglielmo Marconi once said, with some modesty: 'I have striven to give the world improved and cheaper means of communication by means of electrical transmission through space.'

But in the London of 1896 no one knew who Guglielmo Marconi was, but equally no one could have dreamt what he would become and what he would achieve. Marconi's work was destined to completely change the world in just five years. Few other scientific advances or pioneers can claim to have had such an effect.

Being Marconi he tackled all the problems head on with a quiet reserve, but ruthless dedication to the task as he moved from experiment to demonstration, and from test to trials, each carefully calculated step pushing his system, himself and his team harder and harder. Young as Marconi was, his dedication and single-mindedness coupled with his gentlemanly demeanour was very different from the popular Victorian image of the 'mad inventor', and completely different from some of his more illustrious competitors. His continuous successes despite the many obstacles, inspired loyalty in his small workforce of engineers, most of whom had learned their trade in the established Victorian business of telegraph cables.

Guglielmo Marconi gave the first public demonstration of his system of wireless communication in this country in 1896, he then built the world's first permanent 'radio' stations and just five years later his first wireless signal crossed the Atlantic Ocean.

The birth of radio was an explosive process that took just five years, during which the centre for his work, demonstrations and research was always the Isle of Wight.

On that small island off England's south coast Marconi built his reliable system for wireless communication and with it he changed the world.

CHAPTER 1

Prelude to wireless

Act 1
The Ardent Electrician

Guglielmo Marconi was born on the morning of 25th April 1874, at the Palazzo Marescalchi, (today called the *Palazzo Orlandini*), via Asse 1170 (today called *via Tre Novembre 5/7*) in the Italian city of Bologna.

He was the second son of a 'runaway' marriage between Giuseppe Marconi, a wealthy Italian landowner and Irish born Annie Jameson, youngest daughter of Andrew Jameson of Daphne Castle and Fairfield, Enniscorthy, County Wexford. The Jamesons were a well known family of whiskey distillers in the Dublin area and the castle where she grew up was located only a short distance from the brewery. Annie's grandfather was John Jameson, founder of whiskey distillers Jameson & Sons and on the other side of her family she was also related to the Scottish Haig family, also well known Scotch whisky distillers.

The young Guglielmo Marconi had been born into a life of some wealth and privilege, but his upbringing was to prove to be anything but normal. Marconi never received a regular scholastic education as his mother Annie preferred to provide private lessons with private tutors. One of the major obstacles to a traditional course of study was his frequent excursions with Annie. The Villa Griffone at Pontecchio was only 15 km from Bologna with its cold climate and Annie was in the habit of spending long winter periods elsewhere, especially at Livorno and Florence. One of the major attractions of these two cities, apart from the better climate which was beneficial for Annie's health, was that both Livorno and Florence had a significant number of English expatriates at that time.

So much of Guglielmo Marconi's early childhood was spent travelling with his elder brother Alfonso and his mother, while his father enjoyed his role as squire of his country seat.

As a child Marconi was not a success, either academically or socially. From an early age it was clear that he was more interested in devising scientific toys than in any form of school work. He was always building 'gadgets' and dismantling household appliances to see how they worked. He began talking about 'my electricity' and his experiments became more and more involved and consumed increasing amounts of time. Marconi once described himself as simply an *ardent electrician*.

Guglielmo never attended a public school for any period of time. As his mother moved the family to Florence or at Leghorn, Guglielmo had a private tutor in each of the towns. He disliked formal lessons, but was very fond of reading, and would often shut himself away in his father's library for hours reading about the subjects that most interested him. He liked history – particularly Greek history – and also mythology and exploration; but the subject that appealed to him above all others was physical and electrical science. Marconi's favourite sport was fishing and he had the patience to become an expert angler. Next to fishing he liked to ride horseback and travel, especially by sea. Annie did pay careful attention to her son's religious education and always read them two chapters from the Bible every day.

Marconi's mother was 17 years younger than his father and she enjoyed being part of the late Victorian European 'social scene'. As Annie Marconi travelled extensively with her sons, sometimes staying for long periods of time in various European cities, the winter migrations were extremely educational for young Marconi who later became a world traveller himself. It was undoubtedly his mother who gave him the social skills to move easily among different cultures.

But when Annie's sister Elizabeth moved to Leghorn on Italy's upper west coast from Florence, naturally Annie and Guglielmo followed. Giuseppe rented an apartment at Viale Regina Margherita and in 1891 Guglielmo enrolled at the Technical Institute at Leghorn known as the Leghorn Lyceum (Livorno) in *via Cairoli*. While there he attended Professor Giotto Bizzarrini's lectures on physics and Annie always made a great effort to find him the best possible tutors in his favourite subjects, mainly in the sciences and music. It seems that it was during this study period between 1891 and 1892 that he became increasingly interested in experimental science. From the autumn of 1891, for more than one year the now 17 year old Guglielmo Marconi was privately educated by Professor Vincenzo Rosa from the *Liceo Niccolini* in Leghorn.

Villa Griffone, c. 1894

Annie Marconi with Alfonso and his young brother, Guglielmo

Vincenzo Rosa had attended the main university in Turin and obtained a degree in mathematical physics in 1876. He was appointed as a teacher in the school of *Monteleone di Calabria*. In 1879 he started teaching in Reggio Calabria, where he received a visit from the school inspector, the Italian physicist Antonio Roiti (1843-1921) who was impressed by his background and employed him as his assistant in Florence in 1880. From 1887 he taught at the high school in Leghorn and began to devise a system of clocks that were synchronised through an electromagnetic signal transmitted by a clock driver.

In the almost nomadic style that characterised his education Marconi visited Rosa's house for private lessons in physics and chemistry. Rosa's house also contained his private laboratory that contained numerous scientific models and instruments. Marconi learned to use them all and even helped Rosa prepare his lessons for the following day at the *Liceo* (a secondary school considered to be the peak of the Italian upper secondary education system) continually asking questions about technical and scientific arguments. Later Marconi began to follow the teacher to his school and acted as his assistant and helper during his lectures.

At Leghorn, the sea air improved Marconi's health, he partook of his love of sailing and he improved his knowledge of physics by reading technical articles about electricity, many of them delivered by his English relatives. It was at the *Liceo* that Marconi was first introduced to the work of the German physicist Heinrich Hertz who in 1887 had discovered electromagnetic radiation, known by the popular press and the scientific community as 'Hertzian waves'. It was also here that Marconi was introduced to a retired telegrapher, Nello Marchetti, who was losing his eyesight.

The two got along well, and soon Marconi began reading aloud to the older man while the old telegraphist taught him the Morse code and how to use a Morse code key to send messages by cable. Marconi became a proficient Morse code operator and he was to personally operate many of his experimental wireless stations over the next five years. It was on the old telegraphist's Morse code key, sitting on a sunny window sill amongst some potted geraniums that Marconi for the first time tapped out the Morse code letter S. Little did he know that those three dots would shape the rest of his life.

Marconi's big idea was to develop an apparatus that could provide a practical system of wireless telegraphy. It did not take long for him to realise he was confronted with a huge and complex jigsaw puzzle. If Maxwell had defined the picture, and Hertz had painted the picture then it was down to Marconi to patiently start and assemble all the pieces. His advantage was that many of the parts required had already been provided by the world's best theorists and experimenters, but no one had even thought about putting them all together. To complete the picture would still need a visionary with a practical mind to sort, develop and refine each fragment to build a practical system of communication for signalling through space without the use of connecting wires.

Marconi continued to absorb parts of the puzzle. He read everything he could find in English and Italian and even struggled his way through French and German texts as well. One important influence was the weekly semi-professional science magazine *L'Elettricita* to which he had a subscription. From this periodical, he extracted many of the early technical ideas that he needed to carry out his experiments. In particular it gave Marconi sources for ordering materials, especially rare or precious metals that he would patiently file down to fill his new *coherer* designs that were the key component of his receiver. It seems quite plausible that Marconi actually came upon the very idea of wireless communication by reading this magazine. Issue 44 in 1893 contained an article which extolled the importance of electricity and claimed that:

> 'The slow vibrations of the ether would allow the marvellous concept of wireless telegraphy without underwater cables, without any of the expensive installations of our time.'

Marconi became more and more obsessed with the idea of sending Morse messages without wires. Hertz in Germany had achieved a distance of at best sixty metres, Professor Oliver Lodge in England probably less, but neither had followed up on this breakthrough because they saw no practical use for the experiments. Hertz wrote:

> 'It's of no use whatsoever... this is just an experiment that proves Maestro Maxwell was right - we just have these mysterious electromagnetic waves that we cannot see with the naked eye. But they are there.'

In fact, when Hertz's students wondered what use might be made of this marvellous phenomenon, he denied that there was any possibility of a practical application. 'So, what next?' asked one of his students at the University of Bonn. Hertz shrugged. He was a modest man with no pretensions and apparently little ambition. He simply replied: 'Nothing, I guess.'

But by the early 1890s many inventors and scientists were busy reproducing

Hertz's experiments. They quickly reproduced and in some cases slightly improved Hertz's experimental methods, but no one had achieved anything significantly different to Hertz's proven range of transmission, nor attempted to send intelligible messages by Morse code.

During the winter of 1894–95, Marconi had spent most of his time at his work benches in a secure attic laboratory at his family's villa in Pontecchio. The two rooms at the top of the Villa Griffone had once been part of his grandfathers' silk business but they were now set aside for Guglielmo's experiments. While Marconi worked there only two people ever saw the inside the rooms, known as the *stanza dei bachi* or the silkworms room, except for servants who were occasionally allowed in to clean. They were Marconi himself and his mother. In his new secluded laboratory Marconi collected together a wide variety of electrical equipment and locked himself away during every waking hour. The only time the family saw him was when he issued requests for money to buy long lengths of copper wire, batteries, induction coils and other electrical parts.

Guglielmo was delighted to have his own laboratory, and he carefully locked the door whenever he left the room so that no one should be able to find out what he was doing or disturb his arrangements. One day, his cousin Daisy found him in his room cutting strands of very rough wire into lengths of about an inch and a half with the aid of an enormous pair of scissors. In cutting one of these strands of wire, he accidentally cut a piece out of one of his fingers. But he was so engrossed in his work that he appeared to feel no pain, and he refused to stop what he was doing, remarking casually that he would have the finger dealt with later by the chemist. By the time he was eighteen the walls of his room were lined with shelves of mysterious apparatus and jars of different coloured liquids.

More than once his father must have wondered exactly what was going on in the third floor of his home; as boxes full of batteries, transformers, oscillators, bells and wires continually arrived at the Villa. His mother remembered that there were wires everywhere and the room was soon full of spark-gap oscillators, condensers, resonators, coherers, and induction coils.

Then followed one of the most intense periods of his life as Marconi worked day and night, losing interest in everything else. His mother became worried at his drawn, wan face and his eyes always revealed the need of sleep. On very hot days the attic would turn into a roasting oven but Marconi worked on. As he grew thinner his mother became concerned, as a child Marconi hadn't been the fittest of boys but the trays of food she left on the landing outside the attic door were often left uneaten.

Thirty feet was the largest space that Marconi could find inside the building and a few days later signals were transmitted from one end of the house to the other. He then transmitted from the house to the lawn where he was able to ring a bell on the ground floor by pressing a button on the third floor. His mother was firmly convinced that her boy was not playing tricks, but his father was less convinced. His son had a reputation as something of a trickster and if this was some sort of magic trick he would soon figure it out. He suggested that Guglielmo send the Morse code letter 'S' while he would go to the receiver on the lawn, and if the machine tapped off three short dots he would be sure that the magic worked through the air and the building in between. The system worked. It was indeed a kind of magic.

In January 1895 Marconi wrote: 'I at once obtained results, which surprised me, and which I realised were new.' His father was now more willing to lend financial assistance. When prevailed upon by his wife, Giuseppe contributed 5,000 lira (about £900 or today equivalent to £85,000) to his son's experiments and the cause of wireless communication. By spring 1895, with the financial assistance from his father and the moral support of his mother, the young inventor was ready to test his wireless system in open spaces, and if successful, he was determined to offer it to the world. But first Marconi was about to go on holiday.

In the southwest of Switzerland, close to the French border, the small town of Salvan is 934 metres above sea level and lies about 4.5 miles (7km) north-west of the town of Martigny. It was a small community of some 1500 inhabitants and in the last years of the 19th century the village was known as a health resort reputed for its salubrious climate and good air. Still recovering from a respiratory ailment and completely exhausted after a winter of experiments and research it is thought the 21 year old Marconi had been recommended or even instructed to go there by the family doctor. Marconi knew and liked Salvan having been there the year before.

Overlooking the deep Triente gorge at the edge of the Mont Blanc range, access to the village was by a steep and winding path, nicknamed *route de Mont,* leading from Vernayaz in the Rhone valley. Accompanied by mules to carry their luggage and equipment, Marconi, Alfonso and Luigi visited Salvan in July 1895. Alfonso soon left, but Guglielmo Marconi rented accommodation on the second floor of a house in the village on *Millionaire Street* that belonged to the uncle of the boy who was soon to become his assistant and helper in Salvan. Salvan was essentially a pastoral community where villagers were happy to welcome tourists to help improve their meagre standard of living.

With the help of a local teenager called Maurice Gay-Balmaz, Marconi set to work for a month and a half trying to progress his attic experiments. Maurice had quickly become intrigued by some odd-looking metallic apparatus lying in the grass and eagerly agreed to assist Marconi in the experiments that were to follow. Marconi is believed to have located himself on a flat-topped rock called *La Pierre Bergère*, also known as the Shepherdess stone, his small transmitter consisting of a battery, a Ruhmkorff induction coil, a Righi oscillator (spark generator) and an aerial.

As Marconi keyed the spark generator the young Maurice gradually moved further away with a receiver which sounded a bell when a signal was received. Progress was slow, with long hours and many adjustments were required before success was achieved. Maurice eventually carried the receiving aerial raised on an eight foot pole and waved a red flag when the receiver bell sounded and a white flag when no sound was heard. But by the end of July 1895 Marconi had succeeded in transmitting signals over just a few dozen metres.

It was at that point that Marconi made one of his key breakthroughs. He described it when he received the Nobel Prize in 1909:

> 'In August 1895 I hit upon a new arrangement which not only greatly increased the distance over which I could communicate, but also seemed to make the transmission independent from the effects of intervening obstacles.
>
> This arrangement consisted in connecting one terminal of the Hertzian

oscillator, or spark producer to earth and the other terminal to a wire or capacity aerial placed at a height above the ground and in also connecting at the receiving end one terminal of the coherer to earth and the other to an elevated conductor.'

Marconi had raised one wire from his equipment to form an 'aerial' and dropped the other to form an 'earth' connection. Essentially he had created a loosely coupled transformer. There was no theoretical reasoning behind this change; it was just something else to try. But it worked. His system suddenly came to life and started to behave in a predictable manner: 'That was when I first saw a great new way open before me'.

Marconi said later:

'Not a triumph. Triumph was far distant. But I understood in that moment that I was on a good road. My invention had taken life. I had made an important discovery.'

It was a true empiricist's discovery. The young Marconi possessed so little grasp of the physics underlying wireless that he even suspected for a while that he had discovered some new and better form of non-Hertzian waves. But it didn't really matter.

From the *La Pierre Bergère* rock, links were established with the *Rochers du Soir, Ladray, The Cretdu Serré* and even with a piece of land at the top of *les Marecottes*, more than half a kilometre away.

Marconi never recorded or stated when these first crucial tests occurred, only that they occurred in August 1895. By the end of that August, using various apparatus and components and with his new elevated aerial Marconi was able to transmit Morse code over a distance of more than 0.6 mile (1km). In Salvan there is a plaque that reads:

'On this spot in 1895, with local assistance, Guglielmo Marconi carried out some of the first wireless experiments. He first transmitted a signal from this 'Shepherdess Stone' over a few metres and later, following one and a half months of careful adjustments, over a distance of up to one and a half kilometres. This was the beginning of Marconi´s pivotal involvement in wireless radio.'

This was the first time that Marconi had demonstrated his system away from the Villa Griffone estate. His part time assistant, Maurice Gay-Balmaz, became the first independent witness to the birth of modern radio communication.

Marconi spent only six weeks in Salvan, but it had been time well spent. He had moved his system out of the laboratory, packed it up and then rebuilt it in another country with inexperienced helpers. He had refined his technique and his equipment with countless trial and error experiments. He knew his system worked, worked well and more importantly he knew its limits. He returned to Italy as quickly as possible, sure in his own mind that the Italian authorities would now support him. Marconi could now 'transmit' his bell ringing signals over a distance of one mile.

Marconi's laboratory on the top floor of the
Villa Griffone, Italy 1895

Marconi's early equipment in Italy

On his return to Italy the receiving station was carried out to *Cappuccini's* hill at the rear of the family estate, over 1,700 metres from the window of the wireless room on the top floor of the Villa Griffone, so the inventor might keep an eye on the entire distance he hoped to cover. At the receiver site he had told he Alfonso and told him to wave a flag if he saw the coherer's hammer tap out three dots, the Morse code letter 'S'. Marconi touched the telegraph key and immediately his brother Alfonso started waving the white flag.

Later, to prove that he could transmit a signal over natural barriers such as hills and between two posts that were not in sight of each other, Marconi arranged to transmit over Celestini Hill. As he sat at the window of his attic laboratory he watched his brother, the Villa Griffone caretaker *Antonio Marchi* and two estate workers, a farmer named *Mignani* who led the donkey carrying the equipment and a carpenter named *Vornelli* who carried the aerial wire and a long pole cut from a tree, set off across the sun drenched fields in front of the house. Alfonso also carried a shotgun.

Marconi touched the Morse code key, and instantly from beyond the hill, which was about three-quarters of a mile across came the 'salvo'; as success was indicated by a shotgun blast that echoed down the Italian valley.

Giuseppe Marconi was summoned for a repeat performance and still spent some time searching for hidden wires. But when he walked back across the fields to the house, he was convinced that his son had achieved *something* important. Exactly what he had achieved was still unclear to him.

But for his son everything had changed. It was at that moment that the younger Marconi realised that there was no obstacle on the face of the earth or distance that could now stop his wireless system. Marconi's idea had always been to send messages over long distances through the air using Hertz's invisible waves. Nothing in the laws of physics, as then understood, even hinted that such a feat might be possible. In fact to the rest of the scientific world, what he now proposed was the stuff of magic shows and séances, in fact a kind of magic.

Marconi's new goal was a search for distance, but translating what was still a laboratory experiment into a viable and reliable system of long distance communication was to be no easy task.

Act 2
On England's Shores

On the 2nd February 1896 Guglielmo Marconi and his mother Annie arrived in Southampton. His ideas had been presented to the Italian Ministry of Telegraphs but had been firmly rejected. Marconi had carefully carried a locked box all across Europe containing his apparatus, but on his arrival in England the customs house agents had immediately confiscated his equipment, fearing it was a bomb or other *fiendish device* capable of placing the Queen at risk. The customs officials, horrified by the wires, batteries, dials and condensers they found in Marconi's baggage, decided the pair must be dangerous regicides, on their way to blow up Queen Victoria. Indeed, the French President had been assassinated only two years before, by an *Italian*.

Eventually mother and son convinced the officials that they were only harmless visitors with a new invention, but by this time the apparatus had been wrecked, the packing cases were full of torn wires, broken batteries and twisted dials. When they finally figured out the contraption was harmless and wasn't a bomb they returned it to Marconi. Marconi's mother is reputed to have said that her son's idea was a bomb, 'just not the kind you think it is. It won't blow up the world. It will just blow down all its walls.' But it was not an auspicious start to his English adventure as they boarded the train for London.

The two lone figures stood waiting on the freezing cold platform of London's Victoria station, well dressed and surrounded by luggage and travelling trunks. Annie was dressed in a dark suit, laced shoes and was wearing a hat with a large veil. Marconi wore an unusual deer stalker hat, pinched front and back at the crown and a thick fur coat, pulled tight around his thin frame. By his feet he still had his two incongruous but now useless black boxes. The pair had come a long way. They were tired, cold and frustrated at having been delayed in customs for many hours. The hour was late and they still had a long way to go. Standing in the shadow of the great Grosvenor Hotel, the Victoria station yard was one of the great terminal stations for the omnibuses from all parts of the city.

The station bustled with huge volumes of horse drawn cab traffic at all hours, and it never failed to impress any visitor from the Continent. As Marconi caught his first glimpses of the unfamiliar city, the young man, barely 21 years old, could never have dreamt how far this journey would take him.

Their cousin, Colonel Henry Jameson-Davis arrived to meet them from the boat-train. He was in fact a first cousin as Henry was Annie's eldest sister's son, but in reality he had not met the young Guglielmo since he was six years old. To his older (by nine years) and more sophisticated London cousin, Marconi seemed rather frail, young and naïve. Little did he know that his journey across London in a horse drawn Hansom cab to collect the visitors would also change his life forever.

On Wednesday 1st April 1896 Marconi wrote to his father in Italy to tell him that on the day before:

> 'I went to talk to [Engineer] *Price*... who seemed to show extreme interest in my case and told me how he had tried to do what I have achieved using an arrangement different from mine without obtaining any good results. He promised me that, if I wanted to perform experiments, then he would allow me the use of any necessary building belonging to the telegraphic administration... of the United Kingdom.'

His *Price* was actually William Preece, Chief Engineer of the British Post Office. Through his cousins' contacts Marconi had been able to sit down and speak with the most influential man in his field and one of the most eminent engineers in the country. At the age of 63, Preece had found the time to receive into his offices an unknown, foreign amateur inventor who was one third of his age. The meeting changed the course of human history.

Preece's offices were located in the General Post Office west building that also housed the telegraph department. Anyone with a proper introduction from 'a banker or other well-known citizen' could visit the Telegraph Instrument Galleries and see the heart of Britain's massive telegraphic empire. In a room measuring 27,000 square feet stood five hundred telegraphic instruments and their operators. It was the largest telegraph station in the world.

The young Guglielmo Marconi
with his early apparatus for 'Telegraphy without Wires',
shortly after his arrival in England in 1896

William Preece

Henry Jameson-Davis

For his historic meeting with Preece Marconi had two large bags of equipment carried in. He set out his induction coil, spark generator, coherer, and other equipment, but realised that he had forgotten a telegraph Morse code key. One of Preece's assistants, P.R. Mullis, soon found one and together he and Marconi set up the sending and receiving circuits on two tables in Preece's office. The test went well. Preece was surprised and impressed by the young Italian, his system and his ideas.

Marconi, with Preece's support was pleased to provide a private demonstration for the War Office in June 1896 again on the roof of the main Post Office building. As he sent wireless signals from one rooftop to another, the sparks from his transmitter snapped and crackled so loudly as to be audible on the street below and pedestrians passing by stood and stared up. The tests went well.

On 2nd June 1896 Marconi filed his second, revised patent drafted by John Fletcher Moulton, QC, who was a distinguished patent agent of the time. Marconi's Patent No. 12,039/1896 title was: *Improvements in Transmitting Electrical Impulses and Signals and in Apparatus Thereof.* It stated:

> 'Whereas Guglielmo Marconi of 71 Hereford Road Bayswater in the County of Middlesex hath represented unto us that he is in possession of an invitation for Improvements in transmitting electrical impulses and signals and in apparatus therefore, that he is the true and first inventor therefore, and the same is not in use by any other person, to the best of his knowledge and belief.'

The preamble explained: 'According to this invention, electrical actions or manifestations are transmitted through the air, earth or water by means of electrical oscillations of high frequency.'

A letter from Carpmael & Sons, Patent Agents, informed Marconi of the successful acceptance of the specifications of Patent No.12039. It also stated that the acceptance was advertised in the Official Journal of the Patent Office. Marconi's ideas were now public property.

On 27th July 1896 Marconi successfully gave his first *official* British wireless demonstration of his wireless telegraphy system between two Post Office buildings. His transmitter was placed on the roof of the Central Telegraph Office located on Newgate Street, St Martin's Le Grand, where the British Telecom Centre now stands. The receiver was located on the roof of GPO South Building in Carter Lane, sometimes misquoted as the roof of the Savings Bank, somewhat further away in Queen Victoria Street. The distance between the two buildings was around 300 yards, and the experiments were a great success 'despite the presence of several intervening walls'. This was much shorter than the distances he had achieved in Italy, but still impressive to the watching Preece, his engineering team and the gathered officials. When the successful demonstration was over Preece turned to Marconi and publically congratulated him: 'Young man, you have done something truly exceptional. I congratulate you.'

In this diary entry George Kemp, an engineer working for Preece at the Post Office, described the first public experiment in wireless telegraphy:

```
'Our first Experiment on the Roof of the General Post
Office in July 1896 was to note if we got every signal on
the receiver that was sent by the transmitter commutator,
the tapping of the tube and working of the relay were
the only point we could watch, and all went well, water
resistance were used which balanced from 1000 to 3000
ohms and two coal dust or charcoal resistance on terminal
of tapper. We then experimented in the laboratory'.
```

This was a good start. It should be remembered that in July 1896 Marconi was completely unknown. He was a young man with practically no formal education, a foreign inventor whose wireless equipment differed in no basic way from systems and ideas that were already known and had already been demonstrated. His family connections and funding were useful, but he would have to work even harder to make any progress.

In August 1896 Marconi successfully demonstrated his system over half a mile to receiving equipment located on the Thames embankment. He had started to prove to the scientific community of London that his system was worth consideration. Marconi had been hampered by the lack of space in the city so the first full demonstration was planned to take place on Salisbury Plain. The tests were primarily organised for the General Post Office Telegraphy Department and were to be personally supervised by W.H. Preece. The demonstrations started on Wednesday 2nd September 1896. The Post Office engineer in charge was again George Kemp, who transported all the equipment to Salisbury on the day before.

It was a huge gamble for the young Marconi. He had made countless modifications to the equipment since he arrived in England, most of which seemed to improve the system. But he also had no idea if he had compromised the long distance performance that had impressed his father back in Italy. He would now find out, but every move he made, even the slightest adjustment would be scrutinised. Indeed any failure would be publically witnessed by Preece, the GPO engineers and managers, Captain Henry Jackson from the Royal Navy (whose own wireless system was already well developed), his support officers and the War Office. But so would any successes.

For the tests Marconi's transmitting apparatus was housed in a shed they all called 'the bungalow', a temporary building located on high ground near 'Three Mile Hill', near Salisbury in Wiltshire. The receiving apparatus with roughly made copper parabolic reflectors mounted on each side of Marconi's coherer detector was placed on a hand-drawn military cart. The plan was to haul the cart to different locations at different distances from the hill, across Salisbury Plain, while Marconi transmitted every ten minutes.

Marconi's early trials on Salisbury Plain

It was less than a year since Marconi had first rung a bell across the top floor laboratory in the Villa Griffone. Now he was standing on a windswept Salisbury Plain surrounded by Post Office engineers and senior officers from the British Army and Royal Navy.

This was the start of the adventure for unlike all the scientists before him, Marconi fervently believed in the commercial potential of wireless telegraphy to provide a viable and practical system of communication. Marconi had left the laboratory but there was a long way to travel. Kemp wrote in his notebook for Tuesday 8th September:

> 'We worked all the forenoon, shifting our position about 2 miles distant but only got a few taps. We then set it up on the hill by the telegraph wires and got good signals again before Major Carr… the dashes being split a bit.'

One important finding during the first series of tests was that Marconi's reflectors were replaced by aerial wires raised to heights over 150ft (45.7m) above the ground.

As the trials started the first signals came in from a distance of a hundred yards, and during the Salisbury three mile hill tests Marconi successfully transmitted over a range of 1.75 miles (as early as Thursday, 3rd September). Although he later abandoned their use during part of the trials, Marconi also demonstrated the feasibility of directional wireless by use of metal reflectors. Marconi found that he was able to send messages across Salisbury Plain in a more or less definite direction. He mounted the Righi oscillator transmitter in a parabolic copper reflector so that the Hertzian waves were concentrated or focused into a flat spreading beam, in much the same way that a searchlight sends a line or cone of light through the darkness. The outside balls were connected to a high voltage controlled by a Morse key while the two large balls control the length of the wave emitted. Marconi also mounted the coherer receiver in a similar parabolic reflector with the coherer mounted at the focal point of the parabolic reflector.

George Kemp

Righi Oscillator Transmitter

Marconi's Parabolic Transmitter

Marconi's Parabolic Receiver

After the Salisbury Plain demonstrations word started to spread rapidly about the young Italian inventor who had 'invented' wireless telegraphy. Popular newspapers even started to refer to the *Marconi Waves*.

Act 3
Telling the World

On Saturday evening, 12th December 1896, a brougham carriage drawn by a single horse conveyed two men through the gas-lit, rain-glistening streets of the East End of London. One of the passengers, William Preece, was a bearded man in his sixties, the other, who was clutching two black boxes, was in his early twenties. The cab passed along Aldgate and turned on to Commercial Street, coming to a stop outside the elegant façade of Toynbee Hall, the educational and charitable institution in London's East End. The General Post Office's Chief Engineer was presenting an evening lecture there, called Telegraphy without Wires. It was to be the first public presentation of his travelling companion's invention. The young man was Guglielmo Marconi.

Toynbee Town Hall

Toynbee Town Hall Wireless Equipment

Despite his recent publicity, many ordinary people in England had probably never heard of Marconi, but still the Toynbee Hall was packed. It was Preece who talked, as Marconi was not the speaker that night. Preece began the lecture with a brief summary of his own efforts to harness the principle of induction to signal across bodies of water. But tonight, he announced, he would reveal a remarkable discovery made by a young Italian inventor, Guglielmo Marconi.

Ever mindful of who might be in the audience, Marconi had satisfied his need for strict secrecy by concealing his apparatus. He had constructed two boxes and painted them black. In one he installed his transmitter, in the other his receiver, with a bell attached. At the start of the lecture one box was at the podium, the other at the far side of the room. During the lecture Preece operated the transmitter and whenever he created an electric spark, a bell rang on top of a box. But then, on a cue from Preece, Marconi picked up the black box that housed his receiver and walked with it through the lecture hall. Whenever Preece wanted to make a point in his talk, he pushed the key, a long spark cracked and hissed and the bell rang. Preece rang the bell over and over and by now the audience could clearly see that there were no wires trailing behind the young man. No matter where Marconi went the bell rang.

At the end of the lecture Preece introduced Marconi. The young Italian said just a few words to his audience in his impeccable, but slowly enunciated English; 'My name is Guglielmo Marconi and I have just invented the wireless.'

The Daily Chronicle reported that:

> 'What appeared to be just two ordinary boxes were stationed at each end of the room, the current was set in motion at one end and a bell was immediately rung in the other. To show there was no deception Mr Marconi held the receiver and carried it about, the bell ringing whenever the vibrations at the other box were set up.'

This may not sound much today, but it was sensational news back then. So much so that it might have been dismissed as a trick had it not been for the presence and support of Preece, who told the audience that it was hoped that in a short time the transmitter would be able to send messages, possibly over a distance of several miles.

Perhaps the most enthralling part of the exhibition was the reaction of Marconi's astounded audiences to these first transmissions. The demonstration of Marconi's system in Toynbee Hall in 1896 was enough to turn the man into a national celebrity. As sensational as it was, it is probable that few people in the audience in rainy Whitechapel that evening realised that they had witnessed one of the most epoch making discoveries of all time.

In fact much of the British scientific establishment now started to regard Marconi with some suspicion, even distaste. To many people Marconi was a very troubling character, something completely new in the landscape of London society. He openly admitted that his grasp of physical theory was minimal and his command of advanced mathematics non-existent. Guglielmo Marconi was a young man of only twenty-two; he was an Italian citizen with an Irish mother. He had spent nearly eighteen months experimenting with Hertzian waves on his father's estate, and now claimed to be able to send signals over distances of a mile and a half. Indeed Marconi had only a superficial understanding of the underlying physics of Hertzian waves, he was far more interested in achieving practical results rather than in the theoretical background which might lie behind them. In effect, he was what we would now term a youthful entrepreneur, lacking an academic scientific background but driven with great energy, enthusiasm and patience and gifted with a personality which enabled him to penetrate with comparative ease the upper echelons of official departments.

Somehow Marconi ignored the politics and passions, egos and reputations that were swirling around the Hall and London's meeting rooms. He knew that there was still a lot of work to do. Spurred on by his success in London and on Salisbury Plain, he spent the winter months improving the system. He withdrew from all negotiations and further trials, ignored his mail and went back to London to stay with his mother. After a few months Marconi moved to 21 Burlington Road, St Stephens Square and in early 1897 he moved again, to 67 Talbot Road, Westbourne Park in London.

Marconi knew it was time to move on. On 19th March 1897 he started the second series of trials on Salisbury Plain. But this time the demonstration was packed and would be conducted under the auspices of the Post Office and Army and Navy personnel, but also in the full gaze of the newspapers. It was a major occasion discussed for many weeks in scientific circles.

The primary purpose of the trials was to determine the maximum range possible with the revised system design. But again Salisbury Plain was his laboratory and now he risked failure not only in front of the Army and Navy but also in front of the world's press. There was no time to prepare, no practice, no fine tuning and no privacy.

For five days from the 19th to the 24th March 1897 Kemp flew kites and balloons on alternate days. The kites carried up to 200 ft of aerial wire, the balloons 150 feet but despite this the ranges achieved with the balloons were better at 5 and 4.5 miles (against 3.8 miles) possibly due to their better stability.

Marconi had broken every record for wireless communication. When the cavalry officer on horseback galloped up to Marconi from the receiver site and announced that signals were being received over four miles the crowd applauded. Marconi remained calm.

Captain Jackson was one of the first to congratulate Marconi, who simply told him that he was very pleased with the result. Jackson later reported that reliable signals were obtained at 4 miles and the maximum range achieved was 7 miles. Jackson also realised that the key to this leap in distance was the elevation of the aerials. Clearly the bucking kites and balloons were not a practical solution, but he knew that the tall masts on warships were ideal for carrying the new non-directional aerial system.

Dr. F.H. Bowman, a consulting engineer to the British Post Office, now reported that the Marconi system: 'Constitutes the really first successful application of wireless telegraphy.'

In Preece, Marconi had found a stalwart supporter who would fight both his private battles and would also defend him publicly. During his 4th June 1897 lecture on telegraphy without wires, Preece described Marconi's invention in detail and in his conclusion refuted the contention that the Italian inventor had contributed nothing new:

'He has not discovered any new rays; his receiver is based on Branly's coherer. Columbus did not invent the egg but he showed us how to make it stand on end, and Marconi has produced a new electric eye, more delicate than any known instrument and a new system of telegraphy that will reach places hitherto inaccessible'

The press now reported the young Italian's every move with furious excitement and Marconi was besieged with letters from all Europe, America and even Japan. Many offered their services to become Marconi's agent. Their letters remained unanswered. Many others came from women admirers, especially one who wrote that his waves made her feet tickle, although variations of this kind of reaction were never ending. Their letters all suffered the same fate.

Marconi's work had started to make an impact, yet he was still a long way from a commercially presentable or practical communication system. But all the publicity did have another effect. As Marconi had always feared, it woke up the entire scientific community as to what might be possible. Both Professor Lodge and the Russian physicist Alexander Popov started to revisit their earlier work, made easier with the widespread descriptions of Marconi's achievements. The Salisbury Plain experiments in September 1896 and March 1897 had started to prove too many watchers that Marconi's system of signalling using Hertz's waves might indeed be the basis of a practical wireless communication system.

For the young Marconi, keen to develop his experiments and see if he could transmit greater distances, the next logical step was also to attempt transmission over a large body of water. The large expanse of water in the Bristol Channel also offered the advantage of providing convenient 'steps' for progressively increasing the distances to be transmitted. These 'steps' were in the form of the Islands of Flat Holm and Steep Holm, roughly equally positioned between the English and Welsh sides of the channel.

There is much detail available about the trials to cross the Bristol Channel in May 1897, as Preece presented all the results in his lecture at the Royal Institution on 4th June 1897. The tests had been conducted, as usual, in the normal British bad weather conditions and the records speak of people huddled in huts on the beach to shelter from the storms. Details of the Bristol Channel trials also come from detailed notes written by George Kemp, still then a Post Office Engineer seconded to aid and assist Marconi by William Preece.

Kemp had first met Marconi in 1896 during the demonstrations on the roof of the Post Office in St. Martin's le Grand. When Marconi looked over the ornate stone balustrade, he saw a short man, (Kemp was only five feet tall) with thick curly red hair and a handlebar moustache watching him curiously from the pavement below. He caught Marconi's eye and shouted up, 'What are you doing there?' Marconi called back, 'Come on up and I'll show you.'

The onlooker arrived on the roof with such remarkable speed that Marconi believed he had scrambled up the eight storey drainpipe. The moment George Kemp reached the rooftop he knew that he would work for Marconi. He eventually 'signed on' with the fledgling Marconi Company as Marconi's 'first assistant' in November 1897. From that moment Kemp devoted himself entirely to wireless telegraphy, becoming Marconi's inseparable assistant for the next thirty six years.

For the Bristol Channel trials Marconi assembled his transmitter on Flat Holm Island, Kemp delivered the receiving equipment to the cliffs at Lavernock Point on the Welsh coast. A Mr. Williams, who was in charge of the Post Office workshop in Cardiff, had erected two 27m high masts before the tests began. One was on an 18m high cliff at Lavernock Point, near Penarth, and the other on Flat Holm Island. At the top of each mast he fixed a zinc cylinder 0.9m in diameter and 1.8m long; this formed the 'capacity area' – the forerunner of the antenna. He also fitted an earth cable of six heavy bare copper wires which ran over the cliff and down into the sea.

After some struggle, communication by wireless involving nine simple sentences was achieved. Following the initial opening exchange there followed detailed technical messages in both directions indicating each end's equipment settings and receiving sound levels. The distance covered from the Island to the Welsh coast was 3.3 miles, over water.

Marconi then decided that despite the appalling weather he would try and transmit across the entire Bristol Channel. Kemp decided to remain at Lavernock in Wales to operate the transmitter. The receiver party eventually moved to Brean Down, near Weston-super-Mare on the English side of the Channel, probably delivered by boat onto the beach. When the receiver and aerial was finally set up the second phase of the experiment began.

At 2.50 p.m. on Tuesday, May 18th 1897 the receiver party flew their kite and clearly heard Kemp transmitting a series of Morse code V's followed by simple phrases until 4.20 p.m. The engineer reported the successful reception from Brean Down to Flat Holm and onwards to Lavernock Point by Preece's induction system. These test transmissions at 8.7 miles were the longest distances covered by wireless communication at that time anywhere in the world.

Marconi was extremely pleased with the results obtained during the Bristol Channel trials having conclusively proved that his wireless waves would cross water; the system could operate in poor weather and it could be moved easily.

John Gavey reported that the Post Office was impressed with the inherent possibilities of the system. Signals, of some kind had got across the Bristol Channel, although it was clear that, in Gavey's words, 'There was still much to be desired in order to convert crude appliances into good working devices'.

At his lecture Preece told the audience: 'The distance to which signals have been sent is remarkable', and added, 'we have by no means reached the limit'.

On 20th July 1897, while Marconi was in Italy undertaking trials for the Italian Navy at the military port of La Spezia, the Wireless Telegraph and Signal Company was incorporated in England. It had a capitalisation of £100,000 and located at 28 Mark Lane, London, EC3, just a few doors down from Jameson-Davis's own office at number 22. The new Company paid Marconi £15,000 in cash, less the legal fees needed for forming the Company, and £60,000 in stock in £1 shares for all his patents in all countries, except Italy and her dependencies, which rights Marconi reserved for himself. £25,000 was provided as working capital. The remaining 40,000 shares were put on to the stock market. Six of the eight first subscribers were corn factors or corn merchants in one way or the other all were associated with the Jameson family. Marconi's new Company was to be a family syndicate. The initial investments came from his mother, the Davis and Jameson relations and friends, but there was no money from his Italian relatives. The shareholders and Directors were Henry Jameson-Davis as Managing Director, Guglielmo Marconi as Technical Director. Most of the Directors were Irish and from Dublin. Henry W. Allen became the first Company Secretary.

On Saturday 24th July 1897, Henry Jameson-Davis put together what was to become the first press release for the new company, sent from his 82 Mark Lane address.

Re Wireless Telegraph & Signal Company, Ltd

I have pleasure in informing you that the Contracts were all duly signed on Thursday last at this office, and the £15,000 was yesterday duly paid in to the credit of Marconi's Bankers, and the documents transferred to the Solicitors of the Company.

Mr. Marconi, who has been in Italy, wired me from there three days ago saying that he had splendid results from trials at Spezzi [La Spezia] where the smallest apparatus had transmitted telegrams twelve miles. He also said that he would be in London on Monday, when we shall no doubt have full particulars of what has been done.

The invention seems to have been tried now up to 12 miles, so the company can if it likes make further long distance experiments or in fact deal with the invention as it thinks fit.

Marconi had thrown down the gauntlet and firmly stated his intention to build wireless telegraphy into a commercially viable system. After the new Marconi Company was formed there was a flurry of patent applications as scientists and investors sought to regain the lead 'stolen' by Marconi's first patent and the formation of his well financed company.

For Marconi it was now time to move on and establish a permanent wireless station, a base where he could push on with his equipment development and his search for distance and reliability. The decision had already been made to base this on the Isle of Wight, off England's south coast. On 23rd November 1897, just four days after joining the Company, George Kemp was in Southampton at Fay's

Boatyard arranging for the first set of masts for the new station. Their destination was to be Alum Bay, on the west point of the Island, right next to the notorious shipping hazard of the Needles rocks.

This was to become the world's first permanent wireless station.

CHAPTER 2

Alum Bay

The World's First Permanent Wireless Station

The Royal Needles Hotel, Marconi's Alum Bay
Wireless Station, c. 1898

With the support of the British Post Office Marconi's experiments had progressed rapidly. In March 1897, he had been able to transmit five miles across Salisbury Plain, in May 1897, 8.7 miles across the Bristol Channel and in October, he nearly doubled his previous best distance by linking Bath and Salisbury, 34 miles apart.

Now Marconi had a crucial first British patent and had formed a private company, but he also had lost official Post Office support and possibly a large potential customer. William Preece had been told by his superiors that he could no longer work with the young Italian inventor because he was now no longer an independent inventor but now represented a significant commercial company and could even be a possible competitor.

It was clear to Marconi that the most practical use for wireless transmissions would be at sea, where ships still sailed blind and effectively disappeared once they were out of sight of land. It was critical to the future success of Marconi's Company and for wireless telegraphy that Marconi should start formal sea trials as soon as possible. He now had to convince either the Royal Navy or the Merchant Navy that his system was essential for their operations.

Marconi's successful trials to date had proved to his satisfaction that his wireless equipment could be the basis of a viable communication system. He was not convinced that his potential customers were as sure, and he knew that he still had much work to do. In reality his system was still a handmade collection of coils and wires, fragile tubes and sparking balls.

Marconi had become increasingly famous and with the benefit of his patent money and his shares in the new Company, Marconi was now reasonably well off. But the 23 year old nonetheless faced a significant challenge in maintaining the confidence of his new investors and new directors to fund the research necessary to develop and manufacture practical and reliable wireless equipment to market and sell. To avoid losing their patience and being beaten to success by other interests and competitors he had to push hard to re-engineer his equipment and achieve longer ranges.

However he soon realised that to develop his equipment further and conduct formal trials and demonstrations he required a permanent home for his experiments, equipment, tools and personnel. It was time to give up his semi nomadic lifestyle. He needed an operational base where he could live, work, develop, build and produce professional equipment. It was Marconi himself who stated that the first permanent wireless station in the world was built at the Royal Needles Hotel. This was located on the Western tip of the Isle of Wight, on a cliff top high above Alum Bay, and was always known as the Alum Bay Station. In 1883, *Stevens Directory* recorded that:

'The Needles in past ages formed the original headland, but time and the violence of the mighty deep have separated them from the mainland and shaped them into fantastic shapes and ragged figures.

The Lighthouse, which formerly stood on the highest point of the cliff is now located on the westerly of the rocks. During the foggy weather, a bell is constantly rung, the vibration being heard at a distance of five miles.

The 'Royal Alum Bay Needles Hotel' delightfully situated near the beach is fitted up in the most complete style. Furnished apartments may be obtained.

The shores of Alum Bay form one of the most curious and imposing scenes on the coast, the lofty chalk precipices of a pearly hue on the one side and the beautiful variegated and strangely formed cliffs on the other, with the various colours and peculiar nature of the layers of sand, intermixed with red and yellow ochre, fuller's earth and black flints.'

Royal Needles Hotel – Alum Bay c. 1884 and c. 1898

In the summer months, the Island coast teemed with tourists and horse-drawn bathing machines which trundled into the chilly waters of the Channel so that female bathers could enjoy a discrete dip. Coastal steamers ran regularly from the pier at Alum Bay to the resorts of Bournemouth and Swanage on the mainland to the west. The millions of years old cretaceous deposits at Alum Bay also provided the local tourism industry with the colourful sand cliffs and coloured sands which have attracted visitors since the early 18th century.

Royal Needles Hotel
and the Needles

Royal Needles Hotel
at the top of Alum Bay Chine

Alum Bay Wireless Station

Royal Needles Hotel erecting the aerial

In November 1897, the young Italian and his small team of engineers took up the offer of rooms for rent at the Royal Needles Hotel on the cliff top overlooking Alum Bay. It was out of season and Guglielmo Marconi's proposal to live in and experiment from the hotel over the winter months was readily welcomed by the hotel proprietors. For the next two and a half years the world's first permanent wireless station would be operated from the hotel and its grounds. Marconi also set a precedent with his Alum Bay Station, insisting that there was always a convenient hotel to accommodate himself and his staff near any of his Company's wireless stations.

Marconi loved the Isle of Wight and the South Coast of England and was to return many times throughout his life. A keen cyclist, Marconi spent much of his limited spare time exploring the beauty of the West Wight and he developed a deep affection for its heritage, natural beauty and landscape. During the three years that he was partly resident on the Island, Marconi often attended Mass at St. Wilfrid's in Ventnor and at the Weston Manor private chapel at Totland built by the theologian William George Ward, near neighbour and great friend of Tennyson, who thought him among the 'most worthy of mankind, whose living like he would not find again.' He often received hospitality from the Ward family who owned the chapel and adjacent manor house. Squire Granville Ward took a keen interest in Marconi's experiments and he developed a close friendship with Fr. Peter Haythornthwaite, the chaplain at Weston Manor, the home of the Ward family.

On 24th November 1897 Henry Jameson-Davis wrote to Marconi at Alum Bay from the Company's 82 Mark Lane address wishing him well with the construction of the masts, encouraged him to get the best deal possible and not be charged extra for their delivery. In the letter he also enclosed a ten pound note and added that Marconi should try to get the hotel manager to honour his cheques in future, and if the manager needed Company authorisation it would be sent to him. These then were still the very earliest days of the adventure.

The Alum Bay site was ideal for Marconi's experiments as it provided a length of open water across to the mainland roughly equal to the best range his equipment had so far reliably achieved. Now it was time to concentrate on testing his equipment at sea and as a means of communication between ships and the shore. In 1897 the site was also reasonably remote from large centres of population, where electrical interference, especially from tramways had already started to become a problem.

The all iron pier at Alum Bay (now long demolished) was located below the Hotel and had opened in 1889 replacing an earlier wooden one built in 1863. It was basically a long landing stage, about 370 feet long with a cafe at its entrance and a gift shop on the shore. It was a landing point for pleasure steamers from the mainland including Bournemouth and Lymington and from Yarmouth on the Island. It provided Marconi with a very useful private access to the Island and the Royal Needles station above, and suitable moorage to undertake his experimental tug installations.

The Alum Bay Pier

Above the pier and coloured sands sat The Royal Needles Hotel. Marconi had taken over a seaward looking private ground floor sitting room. Some of the working capital of the new Company was used to convert this room and the hotel's billiard room into a laboratory, workshop and wireless station.

While the Alum Bay site was still under construction Marconi had negotiated an arrangement with the London and South-Western Railway Company to charter two of their ferries. These were often referred to as tugboats, probably because in operational use passenger response to sharing the crossing with a virtual travelling farmyard had led to the introduction of small barges to be towed behind the steamer to carry such smelly cargo.

From as early as 1796, ferries have been operating across the Solent, linking the Isle of Wight to the mainland. In the early nineteenth century, the poor road systems encouraged people to travel by sea between Lymington in the New Forest area, and Portsmouth. Originally, steam ferries operated a circular route around Lymington, Yarmouth, Cowes, Ryde and Portsmouth. On 5th April 1830 the first ferry service between Lymington and Yarmouth commenced. Using the wooden steamer *Glasgow*, and operated by Lymington based owners she would then continue onto Cowes, Southampton, Ryde and Portsmouth on various days.

By 1880 railway lines connected to both the Ryde Pier and the Portsmouth Harbour ferry terminals. It was therefore a natural progression for the railway companies to acquire the ferry routes themselves. In 1884 the Lymington service was bought by the London & South Western Railway Company.

The 56-ton 85ft *Solent* built in 1863 was the last wooden ship to join the fleet. Three years later the first iron paddler, the 69 ton 98ft *Mayflower* was delivered from the yard of Marshall Bros, Newcastle. Then, as to this day, the dimensions of the ships were restricted by the shallowness of the Lymington River. The railway company strengthened the service with the considerably larger 130-ton steel paddler *Lymington*, built at a cost of £6,000 in 1893.

It is not known what Marconi paid to charter the steamers, but one price quoted in 1898 to charter a steamer for a private trip from Lymington to Yarmouth was £10.

With the paddle ship *Lymington* providing the public ferry service, Marconi privately employed first the *Solent* and later the *Mayflower* installing wireless telegraphy equipment on both vessels. Marconi saw their use as critical to testing and proving the range and effectiveness of his station at the hotel and to prove the feasibility of using his equipment at sea. During the first weeks of the new stations both spark transmitter, and various receivers were tested on board in all sorts of conditions; often the South Westerly gales were so bad that the operators worked up to their knees in water whilst battened down in the cabin. Weather permitting; the steamers daily followed a triangular course between the piers at Alum Bay, Bournemouth and Swanage whilst signal strengths were noted.

The *'Solent'* (1863) and the *'Mayflower'* (1866)
in Lymington harbour, c. 1897

The main Needles Hotel wireless station had been built quickly, a testament to Marconi's continued determination and drive. On 23rd November Marconi and Kemp had purchased a 78ft lower mast, 53ft top mast and sprit from Fay's shipyard in Southampton for £75. They also bought two 50ft poles for 11/9d [11 shillings and 9 pennies] and two 28ft poles for 3/6d [3 shillings and 6 pennies] from Dolton, Bournes and Dolton, located on Tredegar Wharf at Southampton.

On 24th November they hired a tug from Mr. Harper for £4 and left for Alum Bay taking four riggers and one carpenter from Fay's Yard. Kemp's diary records, that despite very poor weather, within five days all the spars were erected, the mast and sprit supported and stayed. Raising the aerial mast in the hotel grounds had been something of a local event. Despite the bad weather, on 26th November on arrival at Alum Bay they hired one boy and eleven men to carry most of the equipment up the Chine to the Needles Hotel. The same team hauled the heavy spar on to the beach as far as low tide would permit. The lower mast was then hauled up the chine to the top of the cliffs and across the lawn to the hotel, requiring the help of most of the able-bodied men from the local village of Totland. Helping out were Totland men Thomas Reason and his son Rowland Reason, who also took on the role of chauffeurs driving Marconi around the Isle of Wight. The main mast hole was six feet deep and securely wedged and chocked to withstand the numerous gales.

Marconi arrived on 4th December with a van and landau packed with equipment. On 5th December Kemp unpacked all the apparatus and cleaned it; stowed it in a store room and drilled a hole in the centre of the plate glass window which was to be used for the leading-in wire from the aerial. He then fitted up the transmitter and receiver and made two ebonite insulators. In the transmitter room Marconi kept the Hotel's thick pile carpet and ran the aerial wire through the large bay windows. It was usually a stranded conductor of 7/20 copper wire insulated with rubber and tape that ran up to the top of the mast located in the hotel's yard.

To receive the Hertzian waves Marconi looked to the work of the French physicist, Professor Edouard Branly who had developed his famous *coherer* receiver in 1890. By 1892 it had become the first practical instrument for detecting Hertzian waves even though Branly himself did not pursue any type of communication system.

Branly's device consisted of a glass cell containing a granular conductor between two electrodes. These were usually small metal plates or cylinders with wires attached, placed inside each end of a glass tube containing loose zinc or silver particles, usually very fine filings. Basically, the coherer acted like a voltage-controlled switch that closed when a wireless signal was received. Once the tube was mechanical tapped or shaken it essentially reset allowing it to be ready to detect the next signal. Marconi and his engineering team spent thousands of hours developing and refining the coherer design and experimenting to make it more sensitive with every conceivable mixture of metallic dusts and powders to fill the tube.

To generate the Hertzian waves Marconi's simple spark gap transmitter used an induction coil (similar to a ignition coil in a car, essentially a step-up transformer) connected between a wire antenna and ground, with a spark gap across it powered by a collection of batteries. At the Alum Bay station the 10 inch induction coil was powered by a battery of 100 Obach 'M' size cells, the current taken by the coil at 14 volts being between 6 and 9 amperes.

Every time the induction coil pulsed, the antenna would be momentarily charged up to tens (sometimes hundreds) of thousands of volts until the spark gap started to arc over. This acted as a switch, essentially connecting the charged antenna to ground, producing a very brief burst of electromagnetic radiation. To operate the spark transmitters generated very high voltages. To obtain a spark of one cm required 35,000 volts, a five cm spark 100,000 volts. A coil that could produce a spark of 10 inches in air reduced to 6 or 7mm when under load and connected to an aerial. The whole system needed high degrees of insulation and the risk of a lethal shock off the coil, spark or aerial cable was a very real danger.

While the various early systems of spark transmitters worked well enough to prove the concept of wireless telegraphy, the primitive spark gap assemblies used had severe shortcomings. The biggest problem was that the maximum power that could be transmitted was directly determined by how much electrical charge the antenna could hold. Because the capacitance of practical antennas is quite small, the only way to get a reasonable power output was to charge it up to very high voltages. However, this made transmission very difficult in rainy or even damp conditions. Also, it necessitated a wide spark gap, with a very high electrical resistance, with the result that most of the electrical energy was used simply to heat up the air in the spark gap. The spark transmitter was very simple in operation, but it presented significant technical problems mostly due to very large induced electromagnetic field when the spark struck.

Operating the spark transmitter also had somewhat unusual risks attached. A heavy duty signalling key, not an ordinary cable telegraph key had to be used to send Morse code properly as the key contacts could also arc. The operator also had to continually observe the formation and shape of the spark, which should always be bright white, but with a blue tinge and to operate correctly it should crackle noisily. If the spark turned 'furry and blue' and struggled to jump across the gap it was probably a sign of the transmitter's insulation breaking down.

The action of ionizing the gas is quite sudden and violent so that each spark was accompanied by a loud bang, bright light and considerable heat. Each Morse code letter was built up of either a short spark (dot) or a long spark (dash). There is a very distinctive odour given off by a spark gap transmitter, essentially a mixture of ozone and metal vapour from the electrodes combined with hot insulation.

There were other hazards for the operators. As they monitored the quality of the spark they were subjected to continual bursts of intense ultra violet light so they risked developing 'arc eye' or permanent retina damage, just like that experienced by welders.

The maximum transmission speed was around 15 words per minute, but the art of sending intelligible and reliable Morse code via spark transmission was 'to shorten the dots and lengthen the spaces'. The spark had to be watched at all times because it was possible when sending the very quick 'shorts' [i.e. dots] that the spark would not appear. When it didn't, the operator had to re-send the message or a part of the message again instantly, the error being highlighted on tape that the mechanical ink-marker machine continually churned out.

Operators had to learn what a correct spark should look like and the skill was in being able to keep the spark able to send a 'short' [a Morse dot], but actually a very short 'short' with a good thick spark. If the spark thickness and quality remained the same for a 'long' [a dash] then the instrument was in correct adjustment. When transmitting everybody had to stand clear as the voltages in all parts of the system were potentially lethal. If a solid arc ever formed at the nipples of the actual Morse code key or it caught alight when sending a message, it had to be 'extinguished by a sharp puff of air from the mouth.'

Other problems the early operators had to contend with were the need to shut the receiver in its own screened metal box before the Morse code transmitting key could be pressed. When receiving the box had to be opened and it was possible to get the coherer to its most sensitive condition by heating it with a gas lighter, but this had its own hazards with paper tape and varnished cables everywhere. Batteries were capable of giving four hours of operating time, which equated to around 2,000 words at 10 w.p.m. However recharging them was a long and difficult process, universally disliked by the operating staff. While charging, the batteries gave off acrid sulphuric acid fumes and potentially explosive hydrogen gas. Strong acid also often spilled whenever the batteries were being moved or replaced.

At the Alum Bay station Marconi later changed to a larger 18 inch coil which was 80% more powerful than his usual 10 inch induction coil and able to generate very high voltages. The length of spark when the system was operating was about 1cm, which was a much shorter spark than the coil could give (10 to 18 inches long without the aerial connected), but Marconi found that it provided a cleaner signal. Marconi also noted that no care was ever taken to polish the spheres at the place where the spark occurs, as the results seemed decidedly better with dull and blackened spheres than with polished ones. Marconi wrote:

> 'I had one 10 inch spark coil and later an 18 inch one. With the latter I began to meet the trouble of unsatisfactory keying, which later on, with larger currents, was to become such a serious one. I experimented with various types of signalling keys used in the primary circuit of the coil, and with various contacts on these keys, carbon, silver, etc. being tried. The battery for driving the coil was a large one of dry cells. At the commencement we had no indicating instruments, although later I had an ammeter in the primary circuit of the coil, and accumulators for driving the coil.'

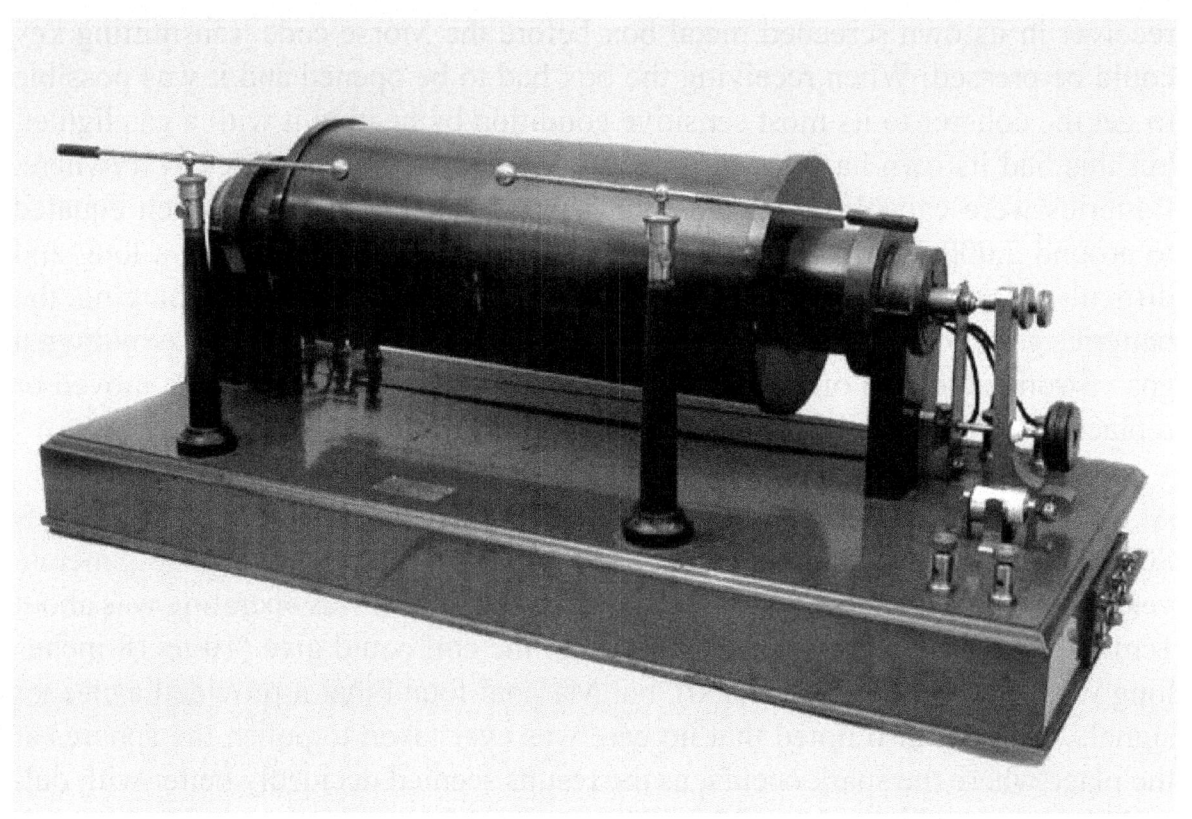

18 inch Ebonite-cased Induction coil from the Alum Bay Wireless Station.
Mounted on a mahogany base with lacquered brass electrodes and a later plaque
from Marconi's Wireless Telegraph Company Ltd. No 886. It measures 42in
(106.5cm) wide.

On 6th December the world's first permanent wireless station at Alum Bay started transmissions. Kemp had fitted a metal cylinder to the end of the 150 feet of aerial wire and hoisted it to the top of the sprit. He then signalled to a coherer receiver located on the Alum Bay pier head below the Hotel and to another located up at the Coastguard Station, about ¾ of a mile away on the Needles headland.

The 120 feet high mast with a wire-netting aerial had been erected in the grounds, without, it seems, giving rise to any complaints from other residents. When Marconi and his engineers were transmitting, the Hotel's guests and visitors were intrigued by the crackle, flashing and hiss of the sparks which generated the mysterious and invisible rays that activated the Morse code ticker-tape machine on the ships.

Marconi soon spread out into various other rooms within the hotel, nearly filling the place in its quiet winter season with pieces of equipment for transmitting and receiving. Hotel rooms became workshops, where coils of wire were wound, wax was melted for insulation, and metals filed down for experimental versions of the receiver's coherer and the whole hotel took on an air of a Victorian factory. In essence that is what the Royal Needles Hotel had become. Marconi soon based his whole team and operations at Alum Bay and we know that one of his earliest recruits, Edwin Glanville was based there from letters written to his stepmother in January and February 1898, until he left for Ireland to help George Kemp with trials for Lloyds of London in July.

Marconi was delighted with the rapid construction and prompt operation of the Needles Hotel wireless station. On its first full day of testing, 7th December 1897, experiments from the station began and despite three weeks of 'atrocious' weather the wireless signals were successfully received at a rate of four words a minute on the receiving ships. Kemp wrote:

```
Dec. 7th - I experimented with the receiver in the tug
and received signals 2½ miles towards Bournemouth and
1¼ miles outside the Needles. The tug steamed back to
Yarmouth having received good signals to within a mile
of the Pier. The tug was obscured by cliffs during this
last trip and, for two minutes before, I transmitted
'Kemp' followed by a number of other names. In the
evening I pasted these signals in a book and these can
still be seen framed on the outside of Senator Marconi's
door at Marconi House. Then I wrote to London and to
Fay & Co. Ltd., from whom I received two coils of rope
6 stakes and 6 thimbles. At night we had a frightful
S.W. gale which made the Hotel tremble and brought down
the sprit owing to the hoisting rope being out by the
metal sides of the sheave at the head of the topmast.
I refitted this with a block the next day and repaired
damages.
```

But on 8th and 9th December the weather was so bad with a 'terrific gale' that the tug could not put to sea and Marconi was taken ill in bed so Kemp had to send for a doctor. The next day Kemp tried again.

> Dec. 10th – Telegraphed for the tug but the weather was still too bad to enable us to go out past the Needles. Made for Milford-on-Sea and received signals at three miles, then we turned and received signals ¼ mile outside the Needles but the tug began to bury herself and filled up the after cabin we therefore had to return, the water in the after cabin being awash with the top of the table where I had the receiver secured. This was the worst day we had so far experienced in the tug

On 14th December testing began again. The plan was that each day, one of the two tugs belonging to the South Western Railway Company, the *Mayflower* or the *Solent*, would steam over a triangular course, from Bournemouth Pier to Swanage Pier and back to the Alum Bay Pier, noting the signal strength as they went.

On 14th December with a 50ft pole fixed to the tug's mast Kemp received signals off Christchurch, but the next day again the weather made sailing impossible. Kemp decided to fit a topgallant mast to give greater height at the Needles Station. He installed the new stays which were of 'the best tarred hemp rope' and fitted a coil of aluminium wire to the top of the aerial. This arrangement was lighter than the metal cylinder offering better storm resistance and it also raised the overall mast height at Alum Bay to 135ft. On December 17th Kemp added a metal cylinder to the tug mast bringing it to 55 ft and signals were received at Swanage

December 18th bought thick fog so again the tug was unable to come. Kemp set about making the chart of the previous days testing and listed the signals sent on the 7th, 10th, 14th and 17th. He also paid the men who had been helping with the past week's work. The weather continued to be poor, but Marconi and Kemp managed to get three more days trials in before Christmas. But, as Kemp was soon to realise, there was no such thing as a Christmas holiday when working with Marconi:

Dec. 20th – The condition of the weather is the same at the Needles. On the tug *The Solent* Mr. Marconi took the receiver while I transmitted to him from the Hotel.

Dec. 21st – The South Western Company could not send 'The Solent' and we had to put up with the smaller tug 'The Mayflower'. With the transmitter on the stern gratings Mr. Marconi received from me when I was 6 ¼ miles out on the triangular course.

Dec. 22nd – I fitted up *The Solent* again as was done on the 17th December, but with the transmitter on the stern gratings and the same one that was used in *The Mayflower* yesterday. Shaped course to Bournemouth Pier, then on to Swanage Pier and back to Alum Bay. Mr. Marconi received from me at a distance of 11 miles on the outward and homeward courses.

Dec. 23rd – I set up all stays and made everything snug at the Needles, and then left for London. I did not remain here long as, on Christmas day, I received a telegram from Mr. Marconi to the effect that he wished me to accompany him to Bournemouth on Boxing Day.

George Kemp later remarked:

'The weather was very changeable, and the tugs did not come every day owing to a south-westerly or dense fog, and often when they did come the weather was so bad and the seas so heavy that the tugs were nearly buried when they turned to alter course. This means that we were confined to the after cabin with either the transmitter or receiver (for we changed the instruments from time to time) and often up to our knees in water.'

In a letter dated 24th December 1897 Marconi wrote:

'I have been busy for the last two weeks carrying out experiments offshore between a ship and the ports of the Isle of Wight. The results have been excellent or I could even say beyond expectation as I could transmit messages during storms and fog to a distance of about 28 kilometres [17miles]. I have been nearly two weeks on board a steam boat during a horrible season but thanks to God everything went well.'

During the course of these experiments, Marconi: 'Had the pleasure of the company and assistance of Captain Kennedy, R.E., who was good enough to draw a map showing the course of the steamer'. Although according to his diary it was actually Kemp who drew the plan.

Before the Alum Bay trials it had been thought that severe weather, with its varying conditions of atmospheric electricity might interfere with or even stop the signals transmitted by the system. Marconi's experience, over fourteen months of continual everyday work brought him to the conclusion that there was no kind of weather which could stop or even seriously interfere with the working of his installation.

The weather for Marconi's sea trials was appallingly bad. The men were tossed about for weeks in the rough seas off Bournemouth, Boscombe, Poole Bay and Swanage, learning about the range and operation of wireless equipment at sea under the worst possible climatic conditions. These early Isle of Wight trials gave an idea of the conditions that Marconi was prepared to tolerate when chasing success and the demands he made on his employees who might have been expected to be rather less dedicated than he was.

On land again, Marconi was able to report momentous results. Firstly, the weather, however foul or impassable did not hinder transmission and signal rates of about four words a minute were achieved. Secondly, he had signalled over the horizon, through a wall of water, at a reassuring distance of eighteen miles and found that the curvature of the earth had not hampered signalling across water.

The equipment was sometimes drenched with sea water, buffeted by high seas and shaken by the continual vibration of the ships' engines. The conditions were so rough that experienced sailors and engineers on the tugs often seemed very anxious about their personal safety. But none of this had any effect on the instruments, which continued to perform reliably and remained in continual communication with Alum Bay. In an interview about the trials, Marconi was now able to strongly refute any suggestion that his equipment was fragile.

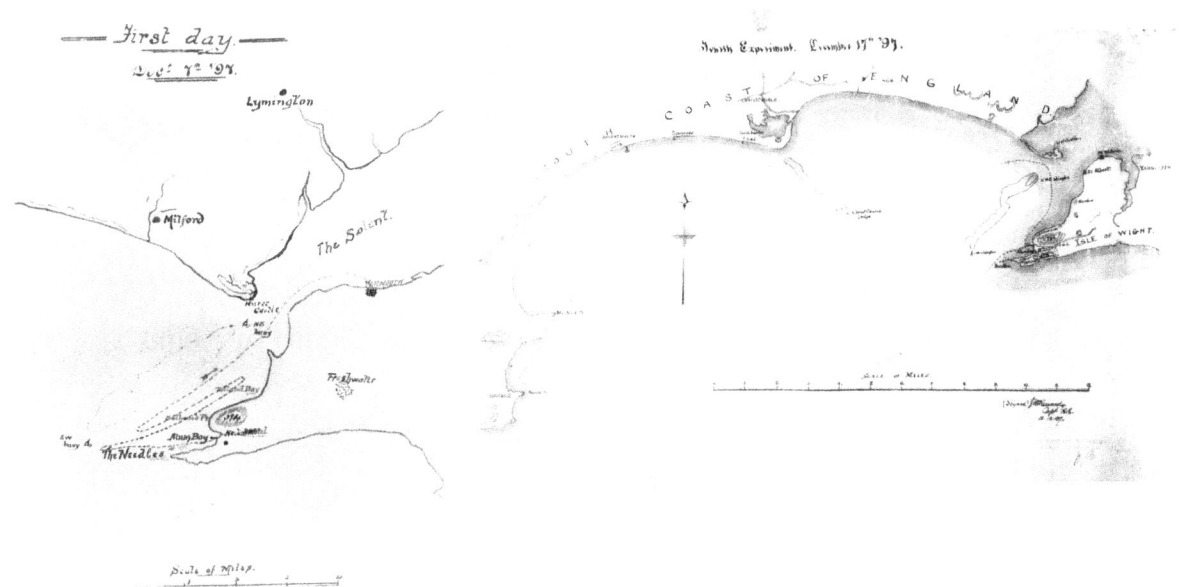

Maps - Alum Bay Station trials. 7th December 1897 and 17th December 1897

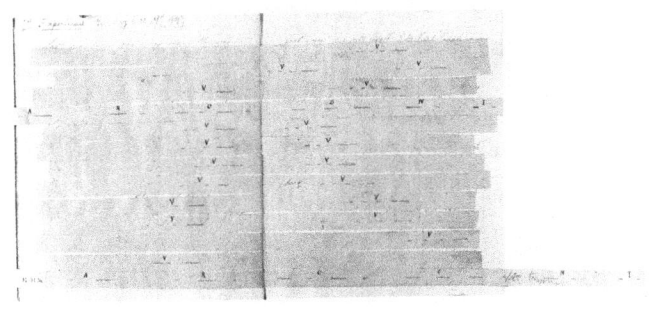

3rd March 1898.
Signed by Marconi

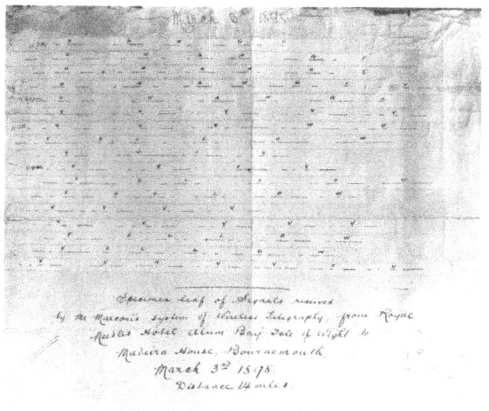

Morse Signal Tapes – Alum Bay.
7th December 1897

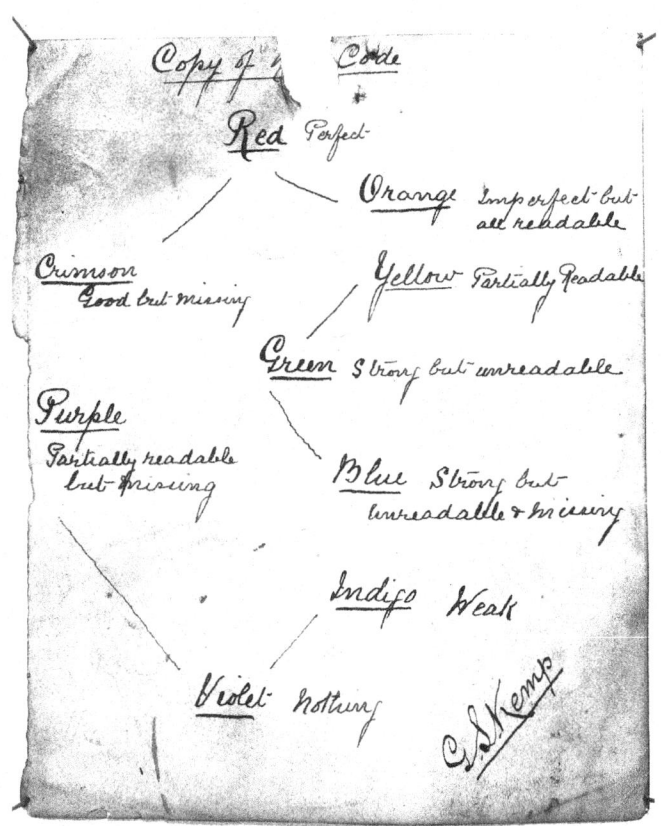

Morse Signal Tapes – Alum Bay. 3rd March 1898. Signed by Kemp

Kemp's original notes on the codes used by the
ship in the Alum Bay trials

Marconi also undertook much of his early fixed aerial design work at the Alum Bay station site. As part of these experiments the wooden aerial mast was often varied in height from 30 ft to 80 ft (although normally it was around 130 feet high) to assess the differences in ranges attained. The sparks that crashed and crackled about the wires were often a source of local entertainment especially in damp weather. During the station's life the aerial system at Alum Bay carried single wires, multiple parallel wires and even 'strip' and 'sausage' aerials made from lengths of rabbit netting. Sometimes temporary additional masts were erected in the hotel grounds, one even borrowed from the Royal Yacht *Britannia*.

Aerial insulation was a constant problem at the Alum Bay Station, especially as the aerial mast, wire and feeder were continually soaked in salt water. To combat this the aerial feeder was bound with rubber tape soaked in rubber solution, but during wet weather and heavy sea mists operation was only possible by washing the window with paraffin oil to prevent water affecting the aerial feed.

Marconi had a remarkable gift for experimentation and an amazing ability to set difficult if not impossible goals, and then get others to passionately believe in his ideas. He was also a good leader, but a hard task master both on his staff and himself, sometimes remaining on duty all night and all day. It was Marconi himself who often helped his team to manufacture the pieces of equipment needed for the experiments and trials and much of the work in these early days was more a matter of improvisation than invention.

Coils were wound and spark gaps adjusted with scientific precision. Yet these early days also required constant application and invention to solve immediate practical problems. Holes still had to be made through a window pane to allow an aerial wire to pass through and smelly cans of wax or varnish for insulation had to be heated up. One constant chore was hand filing metal blocks to produce small piles of metal dust that were used to fill the coherers, the small device at the core of Marconi's receiving equipment that actually detected the weak radio signals.

When the time came for a new series of trials Marconi was always the enthusiastic leader and the most active participant. He seemed to revel in arduous, unpleasant and sometimes even dangerous work even though his senior position would have allowed him to manage from afar. Marconi's personal assistant George Kemp who continually worked with Marconi at the Needles Station later recalled:

'I remember him having to make three attempts to get out past the Needles in a gale before he succeeded. He does not care for storm or rain but keeps pegging away in the most persistent manner.'

Despite the poor weather, progress was always made. Marconi's delight at his ongoing success was evident in a letter he wrote to William Preece on 26th December 1897, Boxing Day:

'I am able to prove repeatedly beyond a doubt that the thickest fog does not in the slightest way diminish the distance to which signals can be transmitted.'

During the course of these trials Marconi set up temporary wireless stations at Fort Victoria about 3.5 miles north on the Island's coast, on Bournemouth pier (14miles) and at Swanage (18miles). By Christmas 1897 Marconi and Kemp were satisfied that they could erect a station at Bournemouth and perhaps even another station at Swanage and keep them in communication by wireless telegraphy day and night throughout the year without interference. Over the next fourteen months the weather, nor 'varying conditions of atmosphere electricity' did not on any occasion interfere with or stop the continual exchange of signals.

Marconi kept his goals and his potential customers in sight. He wrote to William Preece about his results, as one day he hoped that the Post Office might relent and purchase his equipment. Captain Kennedy from the Royal Engineers kept the British Army informed of his progress. For the Royal Navy Commander Hornby and Commander Hugh Evan-Thomas (Later Vice Admiral Sir Hugh Evan-Thomas) visited on 7th May 1898. Evan-Thomas was secretary to an admiralty committee set up to revise the Royal Navy signal book. Afterwards he wrote to Vice-Admiral Sir Compton Domvile, who was chairman of the signals committee, reporting the success of the experiment.

The Royal Navy's annual report for 1898 stated:

'Success at 14½ miles using 10 wpm. Distance directly related to vertical height of aerial; 30' being allowed for 1 mile, 60' for 4 miles and 120' for 16 miles. Height to be achieved using balloons and kites although good vertical heights could be achieved on ships.

Limitations all round when only one frequency [waveband or Tune] was available. Marconi to provide different Tunes so that two [or more] ships can communicate simultaneously with causing interference.

Sparks flying around everywhere and operators receiving nasty shocks. '

Whenever possible, Marconi continued to provide exhibitions and demonstrations at his offices in London, at Alum Bay and his new wireless station installed at the Madeira House Hotel in Bournemouth. Although a lot of 'secret' experimentation went on in the hotel laboratories, Marconi was always willing to risk his reputation and provide very public demonstrations of wireless telegraphy. This above all endeared him to the new popular journals of the day, which had a hunger for exciting and novel discoveries.

Coupled with this, *Signor* Marconi, who was always smartly dressed and available for interview or to provide a non-technical explanation soon became a media star. To the world's newspapers the modest Italian who spoke perfect, if a little accented English, and who appeared to be able to work miracles with a few batteries and a baffling array of wires, was irresistible.

Mr. Joseph Bartlett Garlick, the sub-postmaster at the village of Totland, the closest office to the Alum Bay station, later recalled being invited to test the equipment at the hotel. Asked to send a message he tapped the word '*Marconi*' using the Morse code key connected to the transmitter. To his surprise, a moving bar at the end of the table, not visibly connected, clicked away and repeated the same word in Morse code. J.B Garlick recalled that: 'He [Marconi] picked up the letters on his instrument and was obviously delighted that I had chosen to send his name'.

Sketch of experiment, Alum Bay

Totland Postmaster, Mr J.B Garlick

Later the Totland Post Office and Mr Garlick were to become very busy passing telegraph messages to and from the mainland, at first just confirming successful wireless transmissions and experiments. Later, the Totland Post Office became a communication hub as Marconi organised the birth of his new Company and other trials around the world, but still forced to rely on and pay for use of the wire telegraph. In May 1898 Kemp wrote to Marconi in Bournemouth:

Royal Needles Hotel
Alum Bay
Isle of Wight
11.5.98

G. Marconi Esq.
Madeira Bournemouth

Dear Sir

Eva would like you to send or bring with you a copy of the Electrical Review in lieu of the one you gave to the Offices last Saturday.

I am very sorry this gale is spoiling our cliff experiments but we must make up for it when it is fine enough to launch the coastguard's boat.

I have had to take down the small sprit on the cliff pole and secure the steel wire with 150 ft of netting attached to the bottom of the pole.

I have enclosed leaflets for Mr. Bullocke and hope he is well. I will pay him for the papers and postage when I meet him.

Mr. Bradfield is still in London. I had very good messages from you this morning considering you only had one accumulator and with care we could have kept it up all the afternoon. Two of your messages were perfect and so strong as you could wish. I am now waiting a call up from you as it is past 5pm

I am Dear Sir
Yours Faithfully
G.M. Kemp

Ps We pasted in 10 pages of strips yesterday which had accumulated for the past four days.

By May 1898 interest in Marconi's communication system had grown considerably and a successful and well publicised demonstration was run for Members of Parliament between the House of Commons and St Thomas's Hospital across the River Thames in Lambeth Palace Road. Kemp's diary recorded:

May 18th – Returned to London [from the Needles] taking a 6' coil and fitted up one station at St. Thomas's Hospital in the Treasurer's House and the other in the Smoking room at the House of Commons. Fitted up 2.25ft. Bamboos on the Terrace with a vertical wire having 20 turns of G.P. covered wire in a 9' coil at the top. In the Treasurer's House we had a 25ft. bamboo out of a second storey window and a G.P. covered wire insulated from the top of the lamp-post on the Embankment, 75ft. long on the cliff wire principle. As the rain increased the

capacity increased apparently, and I received from a 1.5
to 1 cm spark. This was at the House of Commons. In the
Treasurer's House the small Righi with ½', 1 millimetre,
½' sparks and we worked the transmitter from 1 p.m. to
7.30 p.m. receiving good messages from the speaker.

May 21ˢᵗ – Packed up all the apparatus and returned it
to Mark Lane.

What the observers found most impressive was that the Marconi men set up their
apparatus in less than an hour. During the demonstration, even the Speaker of the
House sent a message and received a reply. It was a simple event but conducted
with ease, in the heart of the city for some of the most influential men in British
politics.

Another impromptu demonstration came soon after when Mr. Brinton, a Director
of the Donald Currie steam ship line, asked if Marconi's system could report when
a ship passed a shore based signal station. The trial was quickly arranged with the
ship chosen being the 7,262 ton RMS *Carisbrook Castle* on her maiden voyage
(unlike the town on the Isle of Wight which is spelt Carisbrooke). Over and over
again the wireless engineers tried to explain to him how wireless operated, and
the sort of work it could do if given a chance. Finally, with a puzzled frown,
Brinton turned to them. He demanded:

'Do you mean to say that if one of my ships were passing Alum Bay
on its way to port, Mr. Marconi could send word of that fact to shore?'

The engineer replied:

'That is exactly what we mean. The message could be wirelessed from
Alum Bay to Bournemouth, and then telegraphed or telephoned straight
to you here in London. The whole thing would be a matter of minutes.'

Brinton replied: 'I can't believe it. I simply can't believe it.' He stood
silent for a moment and then said:

'Look here. The *Carisbrook Castle* is on its maiden voyage. Naturally we haven't a very good idea of what speed it will make on its first trip. And – well, frankly, we're pretty anxious to have word of it. Do you think Mr. Marconi would be able to let us know when it passed Alum Bay?'

He was told. 'We'll inform him immediately to keep watch for it and send you word the minute it comes in sight.'

RMS *Carisbrook Castle*, 1898

As the ship passed the Needles a message reporting the fact was transmitted over the wireless to Bournemouth from the Alum Bay station and then put on the ordinary wire telegraph lines to Mr. Brinton's office. The telegram read: 'Steamship *Carisbrook Castle*, Donald Currie Line, outward bound, passed the Needles at five minutes past six.'

The Currie Line Director called the Marconi Company's office as soon as the message reached him. The Company's officials assured him that it was no more than the wireless could do at any time. At Alum Bay, no doubt Marconi shook his head. Marconi, to whom the actual message would have been simple, knew the importance of this one small demonstration. Marconi had already developed an uncanny knowledge of just what would impress people, and he utilised it with a deft showmanship that consistently aided the commercial progress of wireless.

Despite being at the somewhat secluded end of the Isle of Wight, Marconi found himself giving numerous demonstrations, including to eminent scientists who often 'dropped in' unexpectedly for a show, but the system always performed perfectly.

The Needles station was becoming quite famous. Marconi's team continued to work testing and demonstrating. Edwin Glanville, a graduate of Trinity College, Dublin with first-class honours in Mathematics and Experimental Science had been one of the first people to join Marconi and had been with him on Salisbury Plain. In a letter to his family on 14th January 1898, Edwin recalled that:

> 'The owner of the Royal Needles Hotel, Mr. Millar was very fond of billiards. Mr. Bullocke [another early Marconi engineering recruit] also plays very well, but Mr. Marconi's play is wonderful.
>
> His theory is that if you hit hard enough you are sure to get something, it doesn't always come off however!'

One guest at the station said that he found something uncanny in the thought that the young man at the key, who seemed as far as possible from a magician or supernatural being, was flinging his words across the waste of sea, over the schooners, over the feeding cormorants to the dim coast of England yonder down the map.

It all seemed so simple, but it was not so easy to teach the world how to do it and perhaps even harder to get them to understand what it could do. Another visitor to the station described the young Italian inventor:

'Standing in the front room of the Royal Needles Hotel, Marconi cut a tall, athletic figure, dark hair, steady grey blue eyes, a resolute mouth and an open forehead. The young Italian inventor. His manner is at once unassuming to a degree, and yet confident. He speaks freely and fully, and quite frankly defines the limits of his own as of all scientists' knowledge as to the mysterious powers of electricity and ether. At his instrument his face shows a suppressed enthusiasm which is a delightful revelation of character. A youth of twenty three, who can, very literally, evoke spirits from the vast depths and despatch them on the wings of the wind, must naturally feel that he had done something very like picking the lock of Nature's laboratory. Signor Marconi listens to the crack-crack of his instrument with some such wondering interest as Aladdin must have displayed on first hearing the voice of the genie who had been called up by the friction of his lamp.'

A journalist wrote:

'It was at the extreme west of the Isle of Wight that I got my first practical notion of how this amazing business works. Here, overhanging the bay is the Needles Hotel, and beside it lifts one of Mr. Marconi's tall masts with braces and halyards to hold it against storm and gale. From the peak hangs down a line of wire that runs through a window into the little sending-room, where we may now see enacted this mystery of talking through the ether. There are two matter-of-fact young men here who have the air of doing something that is altogether simple. One of them stands at a table with some instruments on it, and works a black-handled key up and down. He is saying something to the Poole station, over yonder in England, eighteen miles away.

'Brripp--brripp--brripp--brrrrrr,
Brripp--brripp--brripp--brrrrrr--
Brripp--brrrrrr--brripp. Brripp--brripp!'

So talks the sender with noise and deliberation. It is the Morse code working--ordinary dots and dashes which can be made into letters and words, as everybody knows. With each movement of the key bluish sparks jump an inch between the two brass knobs of the induction coil, the same kind of coil and the same kind of sparks that are familiar in experiments with the Roentgen rays.

For one dot, a single spark jumps; for one dash, there comes a stream of sparks. One knob of the induction coil is connected with the earth, the other with the wire hanging from the mast head. Each spark indicates a certain oscillating impulse from the electrical battery that actuates the coil; each one of these impulses shoots through the aerial wire and from the wire through space by oscillations of the ether, travelling at the speed of light, or seven times around the earth in a second. That is all there is in the sending of these Marconi messages.'

On 3rd June 1898, Marconi was visited at the Alum Bay station by Lord Kelvin, or more correctly William Thomson 1st Baron Kelvin, the famous British mathematician, physicist and electrical telegraph engineer along with his wife, Lady Kelvin. He was accompanied by Lord Hallam Tennyson, son of England's great poet. The Marconi station was practically within sight of Alfred, Lord Tennyson's home at Farringford House above Freshwater Bay.

Marconi duly demonstrated his apparatus for them and answered Lord Kelvin's interested questions. Kelvin had famously been quoted as being initially unimpressed by the concept of sending messages by wireless, once stating that: 'Wireless, is all very well, but I'd rather send a message by a boy on a pony.'

However now, face to face with Marconi and his successful equipment, Kelvin was impressed by the workmanlike and professional atmosphere at Alum Bay. When Lord Kelvin said it would give him pleasure to send messages to several of his friends by wireless, Marconi hastened to assure him that he would be glad to transmit them. 'But I want to pay, you know', the distinguished guest added.

Marconi smiled at the joke and settled down in front of his transmitter. But Lord Kelvin placed a restraining hand on his arm.

'No, I mean it', he insisted. 'One shilling per message, the regular rate of telegraph. I know a few shillings won't purchase much new equipment, but they might help to call attention to the fact that wireless is available to the public as a commercial medium.'

'Don't you think so?' His eyes twinkled.

'He's right', Lord Tennyson nodded. 'You know he is, Marconi.'

Of course he knew. Marconi hesitated for only an instant and then smiled and said, 'Thank you, I do'. Lord Kelvin, an inventor and experienced scientist was a person with whom it was unnecessary to delicately skirt around the importance of publicity and the chance to make a small piece of history. Marconi added, with a mock subservient air:

'That will be one shilling each, sir.
How many messages did you wish to send?'

The three visitors laughed, put their heads together and started writing. Finally they produced handwritten notes, one to be forwarded to Kelvin's chief assistant in the physical laboratory of the University of Glasgow, Dr. Maclean. Another one was for Sir George Stokes at Cambridge, another to Lord Rayleigh, another English physicist, and one to William Preece at the Post Office in London. The message to Cambridge read:

'This is sent commercially paid at Alum Bay for transmission through ether 1 shilling to Bournemouth and thence by postal telegraph 15 pence to Cambridge. – Kelvin.'

The message to the University of Glasgow read:

To Maclean Physical Laboratory University Glasgow

Tell Blyth this is transmitted commercially through ether from Alum Bay to Bournemouth and by postal telegraph thence to Glasgow. Kelvin

The message was handed in at Bournemouth at 2.30 p.m. and received in Glasgow just 50 minutes later.

'Think I'll send one to my nephew at Eton', Lord Tennyson said suddenly.

'Might give him a bit of excitement, you know – and the boys there will be old enough to send wireless messages themselves pretty soon.' The message read:

> 'Sending you message by Marconi's ether telegraph, Alum Bay to Bournemouth, paid commercially, thence by wire; very sorry not to hear you speak your Thackeray tomorrow.'

> Tennyson.

Marconi also tapped this message out on the Morse code key. The messages were all transmitted by wireless across the Solent to his wireless station recently built in Bournemouth and then by land telegraph to their respective destinations.

Lord Kelvin

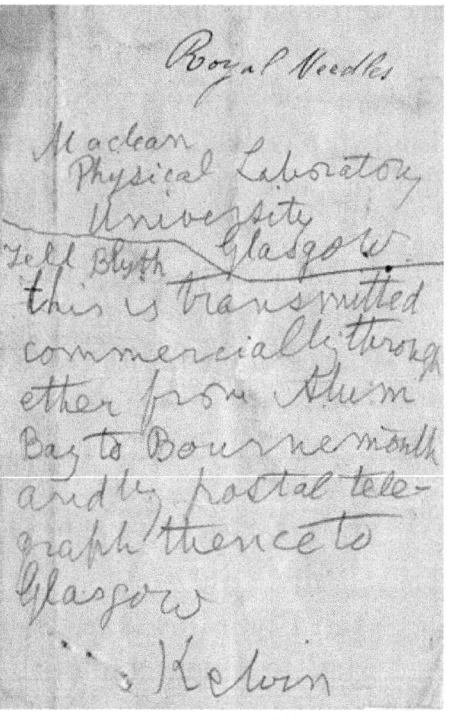

Lord Kelvin's first wireless telegram

Kelvin insisted on paying his one shilling each for what proved to be the first ever commercial radio telegrams to be transmitted. When the visitors had gone the staff at Alum Bay must have smiled at each other. Five shillings was very little, but they knew they had done a good day's business. Lord Kelvin himself had given his approval to a practical new service and was to prove a valuable friend to Marconi. Wireless telegraphy was no longer a mere laboratory toy.

The day after Kelvin's and Tennyson's visit to the Alum Bay station another fortunate event occurred. The stations visitors that day included the Italian Ambassador to England, General Ferrero. When he heard of the previous day's messages he decided to send one himself, a long telegram to an aide-de-camp of King Umberto, in Rome. But when the operator on duty was handed the slip of paper on which the Ambassador had written he looked at it for a moment and then glanced up, puzzled:

'I'm sorry, sir', he said. 'But I'm afraid I don't understand this.'

'It's in Italian', the ambassador said. 'Does that make it impossible to transmit? Must messages be in English?'

'Oh, no, of course not, sir. I'm sorry.' The operator recovered himself quickly.

'The apparatus transmits the letters one at a time – it makes no difference to the machine what words are spelled out.'

A few seconds later the room was alive with leaping and flashing blue sparks, as the operator pressed the Morse code key up and down to produce the dots and dashes of the Morse code. If the receiving operator was confused by the mysterious message he gave no sign. With haste and precision it was sent on over the regular telegraph wires, and within a short time it was received in Italy, exactly as it had been written. The message was between 40 and 50 words long and was entirely in Italian. As this language was not understood by either of Marconi's operators, it could be called a coded transmission. Yet the message was received as perfectly as it was transmitted. After this Marconi could tell the critics who complained of the lack of secrecy in wireless, that messages could be sent in code or cipher, if the senders chose, and thus their privacy would be assured. The wireless could transmit a meaningless jumble of letters as well as the simplest three letter English words.

It was not, he knew, a perfect answer, but Marconi saw these successful demonstrations as a positive indication as to the fitness of his equipment for commercial usage. Before the Ambassador's visit Marconi had already put on a demonstration for both the *Electrical Review* and *The Times,* both of which papers had sent reporters to Alum Bay. They had put the system to every possible test, and, among others, sent a long coded message, which had to be repeated back. In their reports they stated that this was received exactly as sent.

The Ambassador's visit to Marconi had one other purpose and it was a delicate matter. As an Italian subject Marconi had now become liable for military service in his homeland. This seemed to mean that he had only two choices. He could either drop his work and return to Italy and spend three years in uniform or he could choose to become a British subject. The latter course was of course heartily sponsored by his English relations and Company investors. Marconi's work was in a critical phase. Now the world knew the nature of what he was doing and achieving, any delays would mean that other men would carry it forward, perhaps without infringing the Company's patents. This would be disastrous for all concerned.

Unconvinced by Henry Jameson-Davis' recommendation to take British nationality, Marconi turned to General Ferrero, the Italian Ambassador, who was without doubt a very astute politician. Italy's representative in England was also particularly sensitive to any potential shift of power in maritime and navy affairs and was conscious of just how important the science of wireless communication was becoming.

Marconi, by reserving the Italian rights of his patent when his Company was formed had left the door open for Italy to use his invention, and he was rewarded now. The Italian King gave Marconi permission to stay in England. In as much as he owned a boat in Leghorn, His Majesty assigned him, as a naval cadet in training to be an assistant Marine Attaché to the Italian Embassy in London. The irony was not lost on Marconi who less than six years before had been rejected by the Naval Academy at Leghorn. But he was relieved and pleased. With utmost gravity, since he was at no time called upon to perform any duties in his new role, he sent his monthly pay check to an Italian hospital in London. Hence, at this critical point in his career, Marconi was left free to pursue his experiments.

Marconi was quietly encouraged by the support of three men who could be influential and help with his cause. William Preece, the Post Office's Chief Engineer was still helping, albeit now from a distance and Lord Kelvin, who was recognised as the chief scientific authority during that era for wire telegraphy was now convinced. He was also confident that Captain Henry Jackson, himself an early pioneer and experimenter with wireless systems would one day be able to help him place his system with the Royal Navy. These men all believed in Marconi's system. Moreover they all believed in the young Italian. Marconi also now had the support of the Italian Government. His military service fears allayed, he pressed on with renewed enthusiasm.

Kelvin's financial appreciation of the system actually caused the young Italian inventor to break the law, for by accepting the fee he had defied the Post Office monopoly of telegraph services inside the United Kingdom and its territorial waters. It is highly likely that both Marconi and Kelvin knew this when they sent and subsequently advertised the 'Alum Bay telegrams'. Marconi may even have been deliberately looking to provoke an early response from the Post Office. He knew that one day battle this would have to be fought, but on this occasion the Post Office took no action, regarding the payment as a gift by Lord Kelvin to help finance a scientific experiment. Later it was not to be so lenient. Until Lord Kelvin's visit, the Marconi Company had never made a single penny. But five shillings was not really a turning point in the Company's fortunes.

Marconi was to find the search for his first real order an elusive and difficult struggle.

CHAPTER 3

Bournemouth

The Madeira House Hotel

During the construction and set up of the equipment at the Royal Needles Hotel, Marconi had already decided that he required a permanent research and development laboratory to complement the Alum Bay wireless station. Having already operated a successful, but temporary station from the end of Bournemouth pier for part of the early Alum Bay tests, in January 1898 he took up rooms at the Madeira House Hotel, South Cliff, in Bournemouth. On the Bournemouth seafront, near the pier, Marconi now established the world's second permanent wireless station to work with Alum Bay. The Bournemouth station was planned to be a true research and development centre, operating in a controlled environment and located at a convenient distance (18 miles) from Alum Bay.

Bournemouth Pier and Madeira House Wireless Station

Madeira House wireless station and aerial mast

Kemp's diary recalled:

From January 26th, 1897 to January 1st 1898 we were searching for a spot in the vicinity of Bournemouth for making our second Land station, but we found that all the suitable spots could not be obtained. Mr. Jameson-Davis joined us and proposed to risk Madeira House with its narrow strip of lawn extending from the back of the house towards the Bournemouth Pier. As the large lowermast in Fay's Yard was too large for this small garden, I sent for the topmast and sprit to come by road and erected this with a 50ft. pole which I bought locally for a topmast. After a lot of experiments we managed to get good messages each way on January 25th using the two small spheres on the 10' coil rods in lieu of an oscillator which, although necessary for a reflector, was not so good with the plain aerial and earth transmission and reception. The two Brothers Goodbody were at Bournemouth during this test and, as they were Directors, it was nice to be able to do such good work.

The Madeira station started testing from 20th to 25th January and initial contacts across to the Isle of Wight are reputed to have been made using a large zinc cylinder as the receiving aerial standing on a chair in Marconi's hotel bedroom. After some early teething problems, Marconi gave a series of successful demonstrations from the Needles Hotel to Madeira House at the end of January 1898, in front of his Company's Directors and local residents. Lord and Lady Tennyson also attended and took away with them a piece of tape that recorded one of the messages from the Alum Bay station.

Unlike the larger Royal Needles Hotel, space at Madeira House was at a premium and during the summer months the hotel was often full. Kemp noted in his diary that when he returned to the House on April 25th (Marconi's Birthday) that he had to stay in the nearby Dalkeith hotel. On another occasion Alfonso, Marconi's brother also had to stay at the Continental Hotel in South Cliff.

Marconi's new wireless room was only 8 ft by 8 ft and was situated in the basement of the hotel with only a single small window to illuminate it. Next to the wireless room Marconi had a slightly larger workshop some 8 ft by 10 ft that contained an old carpenter's bench with a small metal working vice screwed to one end of it, but again it had only one small window. It did have a good mercury pump for exhausting the sensitive coherer tube detectors, and a hydrogen generating apparatus for making the gas which was required for glass blowing, used in the manufacture of the glass coherer detectors.

In the smaller wireless room Marconi had drilled the window pane to pass the aerial feeder insulated by rubber tube. This was connected to a 10 inch spark induction coil which also had beside it a coherer detector, Morse code inker and Morse code key, all resting on a small table immediately in front of the window.

The only other 'furniture' was a shelf screwed to the wall, on which rested a collection of spark balls, wire and experimental coherers. Underneath the table stood a battery of 100 Obach dry cells necessary to drive the coil and produce the sparks. The aerial was directly coupled to the coil and the wavelength was taken as being four times the length of the aerial wire. The earth connection was provided by a cable that let out to a plate buried in the ground outside the window.

Essentially it was the same equipment as fitted at Alum Bay. With the receivers enclosed in screened iron boxes to prevent local interference, the station was soon successfully operating with Alum Bay on a daily basis. Marconi later wrote about his time at Bournemouth:

> 'Often working in the dark, making apparatus almost entirely by hand, winding coils from wire when the very wire itself was unobtainable in the form in which it was required, so that we had to strand it ourselves before making it into coils – these were some of the difficulties.
>
> At that time there were no moulded insulators on which to wind the coils. Glass, paper and paraffin wax were then the three insulators which I most generally employed. Even now, after nearly a quarter of a century, it is hard to find better substances for the purpose, although many are more convenient in their adaptability.'

A local journalist who visited the station wrote:

> 'His object in Bournemouth is to signal a considerable distance, namely to Swanage on the west and the Isle of Wight on the east or to ships at sea. A flagstaff has been erected and from the top of this a wire runs down to apparatus in the house where instruments for receiving and transmitting messages are placed. The instruments to the ordinary observer are very simple in appearance and can easily be accommodated on a table three feet square.'

It turned out that the new Bournemouth station was quickly going to get busy. By coincidence, in the winter of 1897 the eminent Victorian statesman and former Prime Minister William Ewart Gladstone was terminally ill and Bournemouth had been besieged by the world's press corps. Unfortunately a winter snowstorm in January had hit southern England very hard and had severed all wire based telegraphic and telephonic communication between Bournemouth and London. The reporters were therefore unable to file their latest reports of the statesman's failing health to their newspaper offices in London.

Marconi quickly determined that the wire telegraphic or telephone services via undersea cable were still operational between the Isle of Wight and London. He therefore allowed the press corps to file their reports transmitted by wireless communication from Bournemouth to the Isle of Wight, starting on the 3rd March 1898. These were then passed on to London via the Totland Post Office with great success. This action doubtlessly won him many friends amongst the media, even though Gladstone recovered sufficiently to return to his home, where he eventually died in May 1898.

The 'Gladstone demonstration', especially as it had been conducted in extremely poor weather conditions had been both a fortuitous and very useful publicity stunt. It started to sow the seeds of the possibilities and usefulness of wireless telegraphy.

Towards the end of March, in another raging blizzard, Kemp managed to erect a reflector-type aerial on the roof of Durlston Castle at Swanage over 200 feet above sea level. The 'castle' had been built in 1887-88, designed and purpose built by two local entrepreneurs, George Burt and John Mowlem as a restaurant for the visitors to the Durlston estate. It has fulfilled this role ever since. Kemp's diary records that:

> March 25th – (Great advance) I went to Swanage by train in a snow blizzard taking with me a coil of steel wires and a receiver. I then drove up to the Restaurant, 215 feet above the sea, stretched the steel wire from the gate of Lloyd's Signal Tower to the rocks, in a northerly direction, and from 3.44 p.m. to 4.20 p.m. I laid down in the snow, with my cap over my head and the receiver, and read from the tapper the messages which Mr. Marconi sent to the Needles, one of these messages being Mr. Kemp at Swanage carrying out experiments.

Battling the storms at Swanage, Kemp immediately overheard signals being exchanged between Marconi in Bournemouth and Alum Bay. It not only proved that this was a good position for a station, but that the reflector aerial was as good as or even better than the type used at Alum Bay and Bournemouth. Kemp also successfully lowered a 250 ft wire aerial from the Castle down the cliff face, possibly in an attempt to compare the two aerial systems. Kemp went back to Swanage at the beginning of April.

> April 4th – I went to Swanage and received messages from the Needles Station while those in charge were transmitting with a small transmitter, on a wire leading down to the Pier head. I received all correct, firstly at the top of the glass dome of the Restaurant and, secondly, from the gate of Lloyd's Signal Station as I did on March 25th.

These impromptu experiments were followed by a formal event in April 1898 when messages were sent by various press representatives to the operator at Alum Bay on the Isle of Wight in a private code. The translations of the received messages were later checked and found to be 100% accurate proving beyond doubt that the transmission of messages via wireless was completely reliable.

This text is an early example of the messages received at the Alum Bay Needles Station from Madeira House station on 3rd March 1898, over a distance of 14 miles.

> 'v orange orange v love marconi v v red hurrah orange'.

> 'orange orange orange orange v orange v v v v v v v v v v yellow v v v v v v v yellow yellow'

Marconi later wrote:

> 'The work at Madeira House was some of the most important in the early development of wireless telegraphy, yet the accommodation was very limited, and the apparatus of the simplest nature.'

It was the Madeira House station that received the world's first paid *Marconigrams*, or 'telegrams by wireless' from Lord Kelvin and Lord Tennyson, transmitted from the Isle of Wight on 3rd June 1898. During the rest of 1898 Marconi tended to use the mainland Bournemouth station as the Company's main office and he and members of his staff gave numerous exhibitions of his wireless communication system to practically anyone who cared to visit. Marconi, already courted and chased by the world's press, found himself continually being 'door-stepped' as he entered or left the hotel by a Press Corps eager for the latest news from the magic wireless waves.

Despite the successes achieved at the Madeira House station, there was no more space available. The traditional story is that Marconi's experiments and continual disruption to the Bournemouth Hotel operations and other guests caused problems with the Hotel management but there is no evidence that Marconi fell out with the manager of the Madeira House, Mrs Proctor. It is true that the aerial mast in the garden with its network of guy ropes and wires must have blocked the view of the busy pier, seafront and the sea for the Hotel's holiday makers. At times Marconi's presence at the Hotel led to a virtual scrum at the front doors as press men, visitors and interested spectators all vied for his attention. Besides all this using a spark transmitter, especially in damp weather would have definitely 'entertained' the guests as sparks leapt around anything metal including the bathroom plumbing and taps.

Madeira House also happened to have only a narrow strip of garden facing the Pier so it was not easy to maintain a good aerial height, and 85 feet was the maxium achieved. Marconi was now looking to raise the mast to 150 feet or more which required extra guy ropes and much more space. He also needed more laboratory space and more accommodation for his visitors and growing team. It was time to look elsewhere.

Madeira House, c. 1864

In the centre of this photograph, taken in 1864 is Exeter Rd. In the centre is a house known as *Sandhills,* with smoke coming from the chimney. To the left is Madeira House, site of Marconi's first station in Bournemouth. In Marconi's obituary in the *Bournemouth Times and Directory* on 23rd July 1937 the article suggested that Marconi had moved his research station to the *Sandhills* house after a dispute with the manager at Madeira House. This story has since been repeated, but appears to be a simple confusion with Marconi's move to the Haven Hotel at *Sandbanks* in September 1898.

Madeira House, c. 1906

Looking towards Exeter Road and the South Cliff from the Pier Approach. Second from the right is the Madeira House Hotel, then Sandhills, then Exeter Rd, and finally Cliftonville and South Cliff Hall. The horse drawn bath chairs were a very common form of transport at a time.

CHAPTER 4

By Royal Command

In the summer of 1898 Marconi conducted a series of trials in Ireland including reporting by wireless on the Kingstown yachting Regatta demonstrations in Dublin Bay in July. These first live 'broadcasts' won Marconi tremendous acclaim in the press for what was the first commercial and public tests of his wireless telegraphy system. By the summer of 1898 his exploits and experiments meant that Guglielmo Marconi was also becoming something of a well known local figure on the Isle of Wight. News of the young engineer and his successful tests reached the ears of Queen Victoria who had spent much of that summer at Osborne House on the Island. Osborne House lies less than one mile south-east of East Cowes and had been built for Queen Victoria in 1845-48 at her own expense as a country retreat where her children could enjoy a quiet country life by the sea. After Prince Albert's death in 1861 Queen Victoria spent as much of her time as possible there and she died at the house on 22nd January 1901.

Queen Victoria must have been informed that something intriguing was happening at the other end of the Island, at the Royal Needles Hotel. It is unlikely that she directly requested that a wireless station be installed at Osborne House. According to Marconi it was actually the Lord Lieutenant of the Island who came to meet the young inventor at Alum Bay to ask if it might be possible to establish wireless communication between the Royal Yacht, HMY *Osborne* and the royal residence, Osborne House.

HMY *Osborne* was actually a paddle steamer and part of the Royal Navy. Launched on 19th December 1870 she had been designed by Edward James Reed and had replaced a Royal Yacht of the same name. She displaced 1,850 tons and had been initially used for cruises to foreign countries and later on the short run across the Solent to Osborne House, allowing the Queen to be kept in contact with the political situation in London, delivering dispatches (and on occasion ministers) from Parliament. At Osborne House there was a pier and a landing stage, from which the Queen was able to discretely board the Royal Yacht and come and go as she wished in private, which was impossible at Windsor or in London.

Marconi appeared unconcerned about an invitation from the Queen of England to demonstrate his system. He had after all, already met the King and Queen of Italy. He told an assemblage of his engineers at Alum Bay that he had accepted the invitation:

> 'With true pleasure, for it offered me the opportunity to study and meditate upon new and interesting elements concerning the influence of hills on wireless communication.'

Secretly the young Italian was overjoyed. The Queen wished to be regularly informed of her son's health and progress, as the 57 year old Prince of Wales (later King Edward VII) was spending a period of convalescence. He had suffered a double fracture of his knee cap in a fall down the grand staircase while attending a ball at the Rothschild palace in Paris. On his return to England, the future King Edward VII had retired to the Royal Yacht *Osborne* which had been moored off Cowes for the August regatta in 1898. He preferred to spend his convalescence privately on board the Royal Yacht rather than at Osborne House, where he would continually be under the watchful eye of his mother, Queen Victoria.

In fact Edward was so unwilling to submit to his aged mother's concern for his health that he had the yacht moored in Cowes Bay, two miles away and out of sight of Osborne House. Here the Prince could lead an independent social life, being visited by friends from London, or from the yachts and warships gathered ready for Cowes Week. The Prince could also cruise around the Island, sailing on one occasion as far as Bembridge and the next day to the Needles at the opposite end of the Isle of Wight.

Marconi, ever eager to gain any publicity or support for his system, quickly agreed to set up a wireless station in the grounds of Osborne House. He came straight back from his successful trials and demonstrations in Ireland and installed his equipment at Ladywood Cottage, originally built in 1889 as the Queen's pavilion for the Royal Agricultural Society show at Windsor. After the show it had been dismantled and moved to Osborne and erected at the north east edge of Lady Wood, where it was used by the Royal family as a summer house. Marconi's preparations for the demonstration were as always well planned and the cottage was ideally located at the edge of the estate, on high ground with a clear sea view. At his land station at Ladywood Cottage on the Osborne House grounds he erected a temporary mast 100 feet tall to carry the aerial, connected to a 10 inch coil transmitter.

Ladywood Cottage

Royal Yacht *Osborne*

To the main mast of the Royal Yacht Marconi attached a vertical conductor, raising it to a height of 83 feet above the deck and installed another ten inch coil based transmitter. The aerial wire led down into the saloon, where the instruments were operated and observed with great interest by the various members of the royal family, notably the Duke of York, Princess Louise, and the Prince of Wales himself.

Marconi was now committed, for the first time in his career, to important trials and demonstrations in two parts of the world. Kemp had remained in Ireland after the Kingstown regatta and continued to work with Lloyd's. Marconi was in charge of the demonstrations for the Royal family. He also had operators at both Alum Bay and Bournemouth. His small team was starting to be stretched thin.

The new Royal Island station first transmitted on 3rd August 1898 and was used to keep the Queen fully up to date, maintaining constant and uninterrupted contact for 16 days. Over 150 private communications between the Queen, the Prince and his doctor were handled without fault or error. Many of the messages contained over 150 words with a Morse code transmission speed at around 15 words a minute. Typical of the messages passed were:

4th August. From: DR. FRIPP to SIR JAMES REID

H.R.H. the Prince of Wales has passed another excellent night, and is in very good spirits and health. The knee is most satisfactory.

5th August   From: DR. FRIPP to SIR JAMES REID

H.R.H. the Prince of Wales has passed another excellent night and the knee is in good condition.

On 8th August 1898 the airwaves crackled with:

Very anxious to have cricket match between [HMS] Crescent and Royal Yachts Officers. Please ask the Queen whether she would allow match to be played at Osborne. Crescent goes to Portsmouth, Monday.

Queen Victoria's reply was tapped back across the sea:

The Queen approves of the match between the [HMS] Crescent and Royal Yacht's Officers being played at Osborne.

The following was sent on 10th August by the Prince of Wales while the yacht was steaming at a good rate off Bembridge, seven or eight miles from Osborne:

To the Duke of Connaught.

Will be pleased to see you on board any time this afternoon when the *Osborne* returns.

On 12th August the *Osborne* was steaming 3 miles off the Needles at the other end of the Isle of Wight and maintained constant communication with the Ladywood cottage station. The range achieved was 18 miles of which 14.25 miles were over land, at the time yet another wireless distance record for the inventor. Marconi noted with satisfaction that full communication was also possible when moored in Newton Bay even though the distance was only some seven miles, the hills lying between completely screened both stations even to the tops of their aerial masts. Contact was also fully possible with the Alum Bay station 8.5 miles away despite the presence of Headon Hill and Golden Hill in between. These hills were some 45 feet higher than the Alum Bay aerial, and some 314 ft above the *Osborne's* mast.

Marconi's system worked well despite the ship's aerial being in close proximity to one of the funnels and a large number of wire stays. What seems to have amazed the audience above all else was that messages could be sent at any time, even while the yacht was ploughing through high waves, and even rain or fog did not stop the magic waves.

The Queen is recorded as having been extremely pleased with her regular morning report on the Prince's condition and the Duke of York was so interested in the system that he paid regular visits to the *Osborne's* new 'wireless' room. The Prince of Wales was also extremely pleased with the whole system and presented Marconi with a scarf pin and wished him every success with the invention.

It was around this time that Marconi committed his now famous *faux pas* with the Queen. In fact there are several versions of the story. One popular version has the young Marconi briskly walking through the grounds of Osborne House one morning when he met the Queen who was out 'walking in her bath chair.'

The Queen was seventy-nine, the inventor twenty four, and they were both obdurate people. The lady was addicted to her privacy and had issued orders that it be respected at all times. The inventor, concentrating on his problems, apparently strode through her gardens taking a direct route to his new installation, his mind probably absorbed in the problems which the East Cowes hills might present. 'Good morning Your Majesty'. Marconi said: 'Isn't it a lovely day for a walk?'

This, of course, was not the done thing. A mere mortal, even if their name is

Marconi, does not speak to Royalty, let alone the Queen of England, without being spoken to first. 'Dismiss that young man immediately' she commanded.

An aide replied: 'But that is Marconi, the inventor of wireless.'
The Queen is rumoured to have retorted: 'Well get another electrician!'

Another and perhaps more plausible version is that Marconi used to take a short cut to his hotel through the Queen's gardens. A gardener once stopped him and told him to take another route, but Marconi refused and the Queen was told. 'Who is he? What does he do?' she asked. 'He is experimenting with electricity and wireless signals'. 'Get another electrician' she commanded. She was gently told: 'Alas your Majesty, England has no Marconi.'

We have no way of knowing which story is true, but it was a minor matter. Marconi was invited aboard the Royal Yacht and continuous and reliable contact was established between Osborne House and the station at Alum Bay. This also linked the east and west coast of the Isle of Wight by wireless for the first time.

The Queen even sent a carriage to the hotel to fetch Marconi for a personal audience. Like his own Queen, she wished him success and congratulated him on his work, about which the Prince of Wales had told her. Full use was made of the system by the Royal Family and members of the Cabinet during Cowes Week, the Duke of Connaught and the Duke and Duchess of York being particularly impressed by the new invention.

Marconi's system now had Royal patronage. He wrote home to his father to tell him excitedly of his two weeks with the world's most famous royal family, that Prince Edward had presented him with a fabulous tiepin, and that he was granted an audience with Queen Victoria. However, what excited him most was the discovery that he could keep in touch with a moving ship up to a distance of fourteen miles, his signals apparently penetrating the chalk cliffs of the Isle of Wight. It was a time of triumph for Marconi, but to survive he still needed to impress new customers who would buy his equipment.

The newspapers loved the whole unfolding story of the royal wireless link, none more so than a new popular publication which had gone on sale for the first time in 1896, *The Daily Mail*. A full-page illustration showed Marconi at his wireless set, watched by two fascinated ladies, with his signals careering off along a wavy dotted line to the aerial of the Royal Yacht. Marconi later wrote:

> 'In August, 1898, I was invited to install my wireless system between the Royal Yacht *Osborne* and Osborne House, Isle of Wight. The late Queen Victoria wanted to communicate with the then Prince of Wales, during his cruises in Cowes Bay and the Channel.
>
> When the Prince of Wales sent the first message to the Duke of Connaught, he asked me if that was the very first ever sent from English soil. 'Oh, no, Your Royal Highness', I replied;
>
> 'The first message from the English soil I sent to my aged parents to Italy', and the Prince shrugged his shoulders.'

At the end of August Marconi headed back to Ireland to attend Edward Glanville's funeral after he had been killed in a tragic accident. Marconi now had Royal approval. What he needed most now was a paying customer.

CHAPTER 5

Research
at the
Haven Hotel

Entrance to the Haven Hotel

With the growth of his experiments, space and access had again become a serious issue at the Madeira House station in Bournemouth and Marconi's success at Alum Bay left him in urgent need of a larger research and development station and workshop facilities. In reality the tiny garden at Madeira had always been too small for a large mast, which is why the management had given Marconi a fixed period contract. Marconi also sought more privacy, needed more accommodation for his staff and guests and needed an area that had less electrical interference than the centre of Bournemouth. It was now time to move on.

In September 1898, on his return from Ireland, Marconi moved the mast and packed up all the equipment. The new wireless station was located at the Haven Hotel, Sandbanks, near Poole in Dorset. It first transmitted on 4th October 1898 and was to become the Company's base of operation for the next 25 years. Located just four miles up the coast from Madeira House, the Haven Hotel offered much more space, easier access and was also isolated from the growing electrical interference and passing holiday makers and tourists who flocked to the pier and sea front in the centre of Bournemouth. In the latter years of the 19th century the Sandbanks area consisted of just two hotels, a coastguard station and a few private houses, one of which belonged to Lord Wimborne. In 1899 The Haven Hotel was a very small place. For most of the year the only clients were the local pilots, the ferrymen, a few coastguards and perhaps even a few smugglers. Bill Harvey remembered that his father had been in charge of the boats used by Marconi for his experiments at Sandbanks. 'He were a quiet little chap,' was all he would say; 'didn't like to be disturbed, and us Poole men thought nowt o' him.' ...

Kemp wrote in his diary:

> Sept. 30th – Transported masts on a timber waggon and apparatus in a van to the Haven Hotel, Sandbanks, Parkstone, Dorset and, as we had to erect this mast in the water on the right front of the Hotel, it was necessary to get anchors and stakes driven into the beach, with Pile drivers, at low water.
>
> Oct. 4th – Finished erecting the mast at 4 p.m. and at 6 p.m. I received good reliable messages each way.
>
> Oct. 6th – Bored a hole in the centre of a plate glass window for the wire from the aerial, and the messages were good each way.
>
> Oct. 7th – Tested with the aerial in all directions using

the receiver outside, about 100ft from the mast in all
cases, and the direction at that angle made very little
difference. I then returned to London.

The station initially used the same instruments from Madeira House. The mast
was re-erected on the sand dunes of a barren promontory in front of the Haven
Hotel, six miles from the village of Poole and eighteen miles from the Needles
and the Alum Bay station.

English hotels had always offered the young inventor a place to continue his
experiments and demonstrations, combined with comfortable living and fine
food. Now he had found a place where his mother and older brother Alfonso, as
well as the growing staff of engineers he was gathering around him, could stay
together. Marconi and his mother had no time to enjoy the glamorous social life
of London, but they were able to find some relaxation on the breezy south coast
of England. Marconi wrote that: 'The quarters at the Haven were larger and more
comfortable than those I had had up to that time.'

The Haven Hotel was run by a French couple, Monsieur and Madame Poulain,
who thoroughly enjoyed the fame that the presence of *Monsieur* Marconi and his
experiment brought to their hotel. They celebrated it daily by producing superb
food, their star guest having a special penchant for their special roast chicken.
Monsieur Poulain was described as an 'excellent fellow', who was very keen on
catching mullet.

As at Madeira House the first Haven Hotel station transmitter again consisted of
Marconi's standard 10 inch spark induction coil powered by a battery of 100 'M'
Size Obach cells. The current taken by the coil was recorded in the station's notes
as being between 6 and 9 amps at 14 volts. Test instruments comprised only a
voltmeter, an ammeter and a linesman's detector.

Haven Hotel Station

Haven Hotel Station

Haven Hotel Station, 1900

Haven Hotel Station Interior
Apparatus seen through open window,
1898

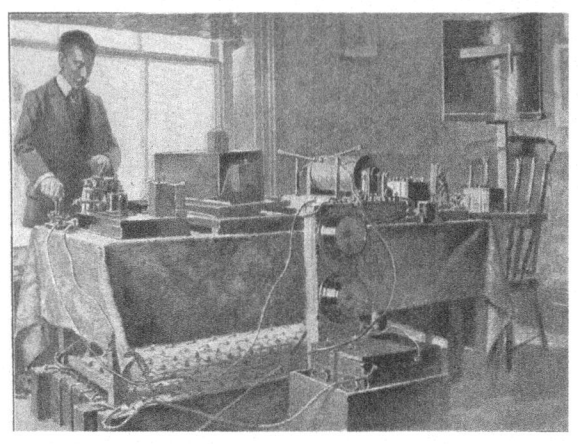

Haven Hotel from the Air

Haven Hotel Equipment
Drawing from McClure's Magazine

Haven Hotel Morse key

Haven Hotel tuned transmitter

The Haven Hotel wireless room was located slightly above ground level with semi-basement rooms under it. Marconi carried out a great deal of experimental work here and a local boatman spoke of a large, east-facing room on the ground floor that served as the main laboratory although he also used some huts in the grounds. He often spotted Marconi through the window standing in front of a 'great flat instrument, several feet square. He looked as though he was playing a piano. Sparks kept flying from this strange machine.'

Marconi's new wireless room was some 18 feet by 18 feet, with two windows. It was as usual necessary to drill one of the windows in order to run the aerial wire through it. One of Marconi's assistants patiently carried this out with no other tools than a bradawl and the point of an old file which he had previously hardened in the sitting room fire. This room was an ordinary hotel sitting room, with a thick carpet on the floor. The team pushed a line of small tables against the wall under the windows. These were used for carrying the transmitter and receiver, as well as a work bench for various experiments.

The tools used for making all this early apparatus were few in number and simple in character. They were principally carpenter's tools, together with a small metal-worker's vice, a home-made hand winding arrangement for making coils, whilst a paraffin wax melting pot supplied the principal form of insulating varnish. In these early days no form of insulating varnish was available, and early trials using shellac were not encouraging. Marconi thought that this was principally due to the presence of water in the methylated spirit which was used as a solvent for the shellac. In general the team returned to the paraffin pot as a quick and ready method of securing good insulation.

Despite all his success, by the end of 1898 Marconi's apparatus was still essentially the same system he had demonstrated two years previously, although step improvements had been incorporated by his empirical, but sometimes random method of 'trying this and trying that'. From the start of his research Marconi had always sought to improve his apparatus by attention to detail, tackling each element of the system mostly by trial and error to find the most efficient form of antenna and by experimenting with various types of coherer.

Marconi had long since realised that there were still three main arguments against the use of wireless telegraphy as a practical communication system. The first was limited range and Marconi had now hit a range limit for reliable communication of around 80 miles, and neither increased power, aerial height or different coherer designs seemed able to push this any further.

With the equipment so far available, Marconi had found that the maximum distance over which he could reliably communicate seemed to increase in proportion to the square of the height of the aerial. It therefore seemed to him that unless he could find some other way, the range of wireless communication was going to be limited by the height to which aerials could be raised. In addition his extensive coherer research was coming to an end and his conclusion was that he knew that this could not offer the quantum leap in range that his system required.

The second problem was that critics pointed out that every wireless message sent was available to everyone and anyone to listen in to, if they had suitable equipment. Even though Marconi had shown that coded transmissions were possible, the perceived inherent lack of privacy in the transmitting/receiving process was holding back potential purchasers and investors.

But beyond both these was the growing problem of interference. Once Marconi had established two permanent stations with large aerial masts at Alum Bay and Bournemouth it became clear that unless a strict transmitting discipline was observed by the stations, if they transmitted at the same time the result was an undecipherable jumble of random Morse code letters and interference. The main cause was that the very spark transmission itself energised a very broad band of frequencies that effectively dissipated its energy. This was considered to be a possible cause of the limited ranges achieved to date, but mostly it meant that any receiving station which happened to lie within range of two or more transmitters would only receive a garbled mixture of all signals which happened to be reaching it.

Whenever the Morse code key was pressed to activate a spark transmitter, the result was a burst of electromagnetic energy produced by the long transmitter spark that made an unmusical note similar to the background atmospherics and actually consisted of a mixture of many waves over a very broad band of frequencies. There was no way to measure wavelength, tuning was unheard of and the actual frequency of transmission was therefore only dependent on the size and configuration of the aerial. The electromagnetic energy leaving the aerial would cover an extremely wide frequency band, so with Marconi's system in 1898 it was chillingly apparent that only one wireless communication channel was possible at a time and hence his quest for a reliable and practical communication system was over. Marconi's system lacked any kind of tuning; his transmitter just blasted out spark-generated pulses of broadband radio noise across a wide frequency range.

The burst of electromagnetic energy from the spark was detected at the receiver, which registered a dot or a dash depending on whether a short or a long pulse of energy was radiated. The receiver consisted of a similar aerial and the use of a coherer which detected the Hertzian waves. A battery operated circuit then operated a telegraph 'inker' which displayed the signal visually on tape. This was a satisfactory communication system as long as there was only a single transmitter. There was again no means of tuning the receiver except to make the aerial the same size as that of the transmitter.

Since his arrival in England Marconi had undertaken many pioneering demonstrations, but the long term development and sales of his equipment was always going to be limited while it remained essentially un-tuned.

If wireless was to be adopted as a general communication system then there would inevitably have to be many transmitters and receivers. It would be vital for each receiver to be able to select any transmitter and receive signals from it and be unaffected by any other transmissions taking place simultaneously. Marconi's stations were now starting to experience significant problems of interference between signals coming from different sources, due to the increasing number of transmitting stations.

At the time every wireless system in the world suffered from the same problems and every inventor and engineer was seeking the solution, a single master invention and hence the master patent that would give them a decisive lead over their competitors.

To tackle the lack of range Marconi at first decided to investigate the possibility of increasing the sensitivity of the receiver. He soon realised that this was a dead end and all worked focused on the new idea of employing 'tuned' circuits in both the transmitter and receiver designs. The principle of tuned circuits was not one of Marconi's ideas. In fact two physicists, Karl Braun in Germany and Oliver Lodge in England had already independently found the answer to tuning.

Lodge called it 'syntony'. He had discovered that a suitable electrical circuit could be made to resonate like a tuning fork and that two circuits tuned to the same frequency could exchange energy. What Lodge did not quite see, however, was how to connect a resonant circuit in such a way that it released its energy as a wave of well defined frequency that only a similarly tuned receiver would respond to.

In Germany, Karl Ferdinand Braun, working independently of Lodge and Marconi had also been trying to determine why it was proving increasingly difficult to increase the distances over which wireless transmissions could be received. Eventually Braun realised that it was the direct coupling of the aerial that was limiting the range of the Marconi designed transmitter. Braun hastily improvised a test of his new 'loose coupled' aerial design on 20th September 1898 and filed a British patent on tuning on 26th January 1899, entitled 'Improvements relating to the transmission of electric telegraph signals without connecting wires'.

The technology of wireless communication systems was about to take a critical next step. Tuning was the solution to the range limitations and the problems of multiple station interference that has dogged Marconi's recent work.

It is tempting to jump to the conclusion that Marconi borrowed some of Lodge's or Braun's ideas and then modified them to suit his own requirements. This, however, is not the case; Lodge's 'syntonic jars' experiment of 1889 did not permit the radiation or transmission of electromagnetic waves for any significant distance. Braun's work remained secret until early 1899. Marconi's experiments with oscillation transformers had begun well before Lodge's patent of 1897 or Braun's in 1899 and well before any constructional details of either system were published.

As with many other areas of technical development it appears that three men were each tackling the same problem at the same time. Wherever it came from, Marconi intuitively seized on tuning as the next step forward and then made it work in practice. That has always been the Marconi way; the laboratory bench and published paper were essential ingredients, but for Marconi any solution always had to be proven in the real world. Being Marconi he revealed nothing about the origins and early development of the new device that would solve the problems. He called his vital modification the *jigger,* a term found in Victorian slang for any sort of gadget.

Its development could have been due to a moment of inspiration or a contribution from any of the increasing band of formally trained engineers he was gathering around him. Marconi began by experimenting with other ways of connecting the coherer, with the principal aim of applying the incoming signal to it as a voltage rather than as a current. In order to do this he introduced a radio frequency transformer into the circuit, with the primary connected between antenna and earth and the tightly-coupled secondary winding connected to the coherer.

The first results were disappointing. Far from improving the sensitivity of the receiver all his experiments found that in fact it considerably reduced it. Other transformers were wound, using different ratios of turns, different wire diameters and different couplings; some improved the receiver sensitivity; others did not.

Between March and December 1897, several hundred transformers of different design were wound and tested, with the Alum Bay station operators working long into the night. Marconi took out patents on three of the most promising ones. The extent of his overall success at this time can be gauged by the gradual increase in the ranges recorded. In later years Marconi in recalling those days was able to say:

> 'The new methods of connection which I adopted in 1898, i.e. connecting the receiver aerial directly to earth instead of to the coherer, and by the introduction of a proper form of oscillation transformer in conjunction with a condenser so as to form a resonator tuned to respond best to waves given out by a given length of aerial wire, were important steps in the right direction.'

His patent, No. 12,326, was applied for on 1st June 1898 for a tuning circuit, mounted inside the receiver to ensure reception only of the signal coming from that transmitting station, ignoring all other signals present in the aerial. Marconi developed and built his first receiving jiggers, or oscillation transformers in the workshop at the Haven Hotel These were coils of wire often wound on paper tube formers. As the jiggers became larger they required larger formers to wind them on. Glass bottles suited the purpose admirably, and formed excellent cores for these early coils.

The circuits were tuned by use of various types of sliding plate condensers, known as *Billi* condensers. It its first form they consisted of a paper tube, built up of thin paper and paraffin wax, with a copper foil coating outside that, whilst inside the first tube another was made to slide and to carry with it a second cylinder of copper foil. These first adjustable-type condensers, developed for use with tuned wireless receivers, enabled sharp tuning to be employed in the receiving instrument.

But Marconi and his engineers soon realised that receiver tuning was only half the problem and the real problem lay with the spark transmitter's inefficient dissipation of the radiated energy as it spread over an extremely wide band of frequencies. The key goal had to be to effectively tune the transmitter. This would significantly increase the transmission distance possible, since it concentrated radiation energy in a small frequency range rather than spreading it out across the whole band.

The reception jigger was followed by a jigger for transmission. Marconi's transmitting jigger was essentially a high frequency transformer, whose primary circuit is connected to a capacitor, forming a resonant circuit which reduces the bandwidth of the signal to be transmitted coming from an induction coil. The signal then goes to the central coil (secondary circuit) whose terminals are connected to the transmitting aerial to be sent into the space.

His early research with tuned and coupled circuits extended over a considerable time, but allowed Marconi to considerably extend the working range of his wireless system. The problem was not as yet fully solved however, for whilst a degree of selectivity was now present, a receiver placed equidistant from two transmitting stations was still unable to separate the two sets of messages.

Marconi looked hard at a further patent taken out by Oliver Lodge in 1898 (Number 29069) but was unable to get any satisfactory results. He tried large metal plates as antennas at the transmitter and receiver, but abandoned these because of the physical impracticability of their use on board ship in high winds. A step forward was made when a vertical wire, earthed at its base, was placed near the vertical radiator; the improvement effected by this gave encouragement to carry out further experimental work. It was to be a long battle.

Shortly after the successful Salisbury Plain to Bath tests in 1897 an engineer named Henry Melville Dowsett had met Marconi at what he later referred to as a 'garden party' demonstration of wireless over a short distance in North London. He was subsequently hired by the Company and summoned to Poole by the Managing Director, Henry Jameson-Davis as the formation of Marconi's Company meant that the Post Office had ceased to supply engineering assistants. As the train from London went no farther than Bournemouth, the new engineer had a long drive on to Poole, much of it through dense pine wood that ended in a mile and a half of sandbank.

H.M. Dowsett eventually found the Haven Hotel with Marconi's laboratory on the ground floor of the hotel, in what had once been a sitting-room. Still thickly carpeted, but with no furniture except for a few chairs, a small table, and benches along the wall under the windows where instruments were arranged. On the day he arrived, Dowsett found two mechanics making coherers and a third winding choke coils. The new man introduced himself to his boss, who smiled amiably but continued what he was doing. Dowsett hung around feeling somewhat ill at ease while everyone went on working. Finally Marconi handed him a piece of metal no larger than a shilling and the oldest and smoothest of files, telling him to make some metal filings for a coherer.

For half an hour Dowsett filed away furiously, increasingly convinced that the whole thing was a leg-pull for the 'new boy.' He accumulated a minute heap of dust, out of all proportion to the effort he had expended and was surprised when Marconi told him he had enough to fill a coherer. Only a fine, clogged-up file, it seemed, produced particles minute enough to serve this purpose. The Poulains' establishment was described by Dowsett as 'modern and comfortable, noted by tourists and yachtsmen for its cuisine and wines.'

Like the Royal Needles Hotel at Alum Bay, the Haven Hotel also soon became a curious mixture of workshop, laboratories and accommodation for the scientific and operational staff and some of their wives. By the turn of the century Marconi and his team had all but taken over the whole establishment, but ordinary holiday makers still came to the pleasant little hotel near the beach. They were no doubt surprised by the huge aerial mast and Marconi's equipment spread out in the hotel as wireless communication was still very much a novelty.

Located on top of a cliff, the words HAVEN HOTEL emblazoned over the seaside entrance could be seen a long way off. Only a low stone wall and a thick hedge of exceedingly prickly evergreen shrubs separated the small hotel garden from the beach, where the hotel had a bathhouse and Marconi had his aerial mast.

Poole was like a thousand other English bathing resorts along the coast, where holiday makers came to enjoy the sea air and quiet bathing. Only one thing set the Haven Hotel apart. Dowsett noted: 'The Marconi three-mast set, 110 feet high, on the sandy foreshore with some of its stay anchors likely to be under water at spring tides.'

The move to Poole now meant that Marconi's equipment was 18 miles from Alum Bay, but no detrimental effect in signal strength was found. If anything reception improved despite a reduction in size of the aerial mast and over 1,000 words were soon being transmitted daily. The aerial mast at the Bournemouth Madeira House site had been extended for a short time to 150 ft high, but had been reduced at Poole to only 110 ft by the removal of an unstable top section. Both stations used the same aerial wire, stranded 7/20 copper wire insulated with India rubber and tape. Later improvements allowed a further reduction in size to just 80 ft, until August 1904 when a 158 ft permanent mast was erected that remained until 1913 when the hotel was extended.

For the past two years Marconi's continual demonstrations of some of the possible applications of wireless telegraphy had given the inventor and his Company a great deal of free publicity. However in private the various trials organised throughout the country and sometimes overseas had exposed a weakness in the core organisation of the new Company. Until November 1898 Marconi was the only real technical authority in the Company; he was also the designer, inventor, engineer and the very public face of the Company. His investigations and demonstrations to date had been undertaken with the support of just one trusted technician, George Kemp, and a few talented engineers including W.W. Bradfield and Edward Glanville before he was tragically killed in an accident in Ireland. The rest of the staff operated the Alum Bay and Bournemouth stations but with public demonstrations and ongoing research work the team was stretched very thin. Marconi was also acutely aware of the urgent need to keep ahead of his competitors. He knew that to achieve this objective he now urgently required more staff with a greater technical and scientific training than his own. He particularly needed to recruit a senior physicist who actually understood the science of wireless. His job would be to progress the experimental aspects of wireless telegraphy but also have enough gravitas to refute or answer the growing tide of scientific scepticism over his claims and progress.

In November 1898 Dr. James R. Erskine-Murray, assistant professor of physics at Heriot-Watt College, Edinburgh resigned his position and became Marconi's principal experimental assistant. From this date Marconi actively began to surround himself with a group of very able engineers and scientists and employed them for many years.

Erskine-Murray immediately went to work at the Haven Hotel and gave numerous wireless demonstrations to promote the new communication system. The hotel also became a mecca for wireless researchers, with an excellent atmosphere and spirit of adventure. A visitor to the station in 1899 recorded that he found in the main laboratory two early Company employees, the brothers J and R.F Cave who were making coherers and P.W. Paget who was winding receiver chokes. Marconi sat at his own table busily fitting 'V' Gap plugs into an experimental coherer. Nearby on another table sat a 500 volt battery used by Marconi for tests with experimental glow discharge coherers. Outside along the foreshore Dr Erskine-Murray and two assistants were conducting parabolic mirror tests using centimetric wavelengths.

During the day the Haven Hotel's other guests were continually entertained by the crackling sparks from Marconi's system, but in the evening the hotel took on an informal air. The Hotel's large parlour had a large fireplace and a cosy dining room. Marconi, his mother and his brother Alfonso, together with Dr and Mrs Erskine-Murray, the rest of the staff and any visitors who happened by, all shared the same dinner table. In the evening Murray would often play his cello, Alfonso his violin with Marconi accompanying on the piano. The trio would play popular classical pieces as the prevailing south-west wind rattled the hotel windows. Annie Marconi often stayed to 'look after' her sons, and those evenings in Poole she recalled were among the most delicious and poignant of her life.

There is no question that Annie provided much of the initial inspiration and strong moral support and encouragement that Marconi needed. How she became so convinced of her son's invention's real worth is difficult to understand as she had no technical background. Like many others before and after, she must have been convinced simply by Marconi's incredible persistence and determination.

For the first few years of Marconi's career, Annie lived in England with the utmost frugality. Guglielmo put Alfonso on a small expense account so that he could buy shares in the Company. One commentator wrote: 'I knew his brother Alfonso. He had none of Guglielmo's characteristics. He was a pleasant, amiable chap; a good natured man, but you would never suspect they were brothers physically or mentally. Alfonso, however, was for years a Director of all the principal companies of the Marconi organization including the American International Marine Communication Company. He died on 25th April 1936 in London.

Marconi looked upon money as simply the unit of reward for all his hard work and had no intention of sharing the fate of early inventors who lost everything. There was a potential conflict between Marconi the entrepreneur and Marconi the scientist, but for the moment he saw no conflict, he liked both roles, since one served the other, and to forward his experiments it was essential that the Company must start to make some money.

Over the next years Annie was to see Guglielmo less and less as he pursued with steely determination his ambition to transmit wireless signals further and further across the sea. For the time being, however, the fame and success he had already achieved was a vindication of her faith in him.

It was clear that Marconi loved the Poole station which for twenty-five years became the field headquarters for Marconi. He was there so often that his mother moved down from London and one commentator remarked that Marconi was most content when he was busy at the Poole station working with Dr. Erskine-Murray.

The Haven Hotel was the one place that Marconi could return to from time to time to relax. Here he met and was wined and dined by distinguished visitors to the south coast resorts and he became a frequent and much admired guest of a wealthy Dutch couple, Charles and Florence van Raalte, who had made a fortune from cigar manufacturing in Holland. In 1901 they had bought Brownsea Island, which lies just off Poole harbour. Over the centuries the island had had many owners, most of whom lost money trying to exploit it in one way or another. The van Raaltes, however, acquired it purely as a private playground for their own pleasure and amusement, and to entertain their upper crust friends and celebrities of the day. The new owners established a nine hole golf course and there was much shooting, with 2,000 pheasants bred each year. Most new male employees had to be musicians and were required to play in the Estate Band. Mr. van Raalte employed 71 servants and workers including a professional golfer and 12 crew members for his two steam yachts. He kept a four-in-hand carriage at Sandbanks. In 1907 Robert Baden-Powell held his first scout camp on the island while it was owned by the van Raaltes.

When he was staying at the Haven Hotel, Marconi often made the short crossing to Brownsea to be entertained by the van Raaltes in their newly refurbished Brownsea Castle, and he became a favourite guest with their teenage daughters. The place may have reminded him of the Villa Griffone back in Italy, where he spent much of his childhood.

One of the van Raaltes' daughters, Margarite, recalled years later the delights of the Haven Hotel and how enchanted Marconi was with Brownsea Island when he was invited to come over and stay. The daughter of Mr and Mrs Poulain, who ran the Haven Hotel, remembered:

> 'The cooking was most excellent: and here, rooms and a big workshop were permanently kept by Marconi, the inventor of wireless telegraphy... My brother, a born mechanic and very much of his generation, was madly interested and I fascinated, by discovery after discovery and at the development of each new invention. All the van Raaltes' guests were given nicknames such as 'Poops' or 'Winkle', and before long Marconi was referred to affectionately as *Marky*.'

Marconi wrote of the Haven Hotel:

> 'My earliest experiments had been on very short waves, and reflectors had been used with these waves; now with larger aerials the wavelength

had increased, but at the Haven I again experimented with short waves, using as an aerial a metal can elevated a few feet from the ground. Fairly satisfactory results were obtained, but difficulty was experienced in getting sufficient power into the circuits, so that once more short waves and reflectors had to be put aside for redevelopment at a later date. It was through the use of jiggers with iron cores that the first magnetic detector came into being, and it was at the Haven that it was made.

Here also, for the first time, I sent two different messages in two different languages simultaneously from one and the same aerial; it was also at the Haven that duplex was first successfully accomplished.'

It was from the Haven station, in constant communication with the Alum Bay station, that Marconi first demonstrated to his satisfaction and to the Company's Board of Directors that high power signals would not swamp the low power sets in use by many ships at that time. His continual developmental success at Alum Bay and the Haven Hotel convinced him that the range, reliability, selectivity and robustness of his system in all weathers were now sufficient for the next step.

Marconi now believed that he had all the pieces to build and execute his grand plan. He wanted to build a permanent high power transmitting station that could generate sufficient power to cross the Atlantic. Marconi also knew his Atlantic project was fraught with daring and might be a little too much for the public mind to grasp.

He also realised the need to avoid premature announcements; all wisdom called for secrecy, so that if the dream failed it would be a matter of disappointment only to the dreamer. If it succeeded, then of course it would be a massive step forward for mankind.

So he set to work on his dream quietly, unassumingly and in secret. Most of all he needed funding. The transatlantic dream would cost a fortune and he would have to reveal his plans to the Company's Board of Directors and get their approval to spend the Company's already shrinking cash reserves.

To realise his transatlantic dream Marconi knew that his equipment needed further development and more testing. His next opportunity soon came from a key potential customer, Trinity House. Marconi, or at least George Kemp, was heading back to sea.

The year 1898 had been a busy one for Marconi with tests and demonstrations taking place in many parts of the country. Marconi had taken his system to Rathlin Island and the Kingstown Regatta in Ireland and worked for Queen Victoria at Osborne House. They had all been triumphs for Marconi and his system. He now had his new Haven Hotel station fully operational in Poole working daily with Alum Bay, but despite all his success, publicity, demonstrations, tests, trials and even Royal approval the Company's order books were deplorably empty.

But the British Army, the Royal Navy, Lloyd's of London, the General Post Office nor any commercial shipping interests had yet placed a single contract for wireless equipment. France and America continued to show great interest in his progress and results and continually asked to organising trials in their countries, but again no contracts or funding was forthcoming. The Italian Navy were happy with the four sets they had bought and *might* buy more, but promises were not going to keep the Company afloat.

Marconi had always envisaged that his system would revolutionise safety at sea, and a prominent part of this was a need to develop a method of reliable communication with lightships. A lightship (or light vessel) is a ship, moored in a fixed location, which acts as a lighthouse and they are used in waters that are too deep or otherwise unsuitable for lighthouse construction.

Communication with lightships had always proved to be a major problem for Trinity House and the Company had spent years trying to connect their lightships and lighthouses to their shore stations by cables. This was extremely expensive and required special heavy duty mountings and fittings that were troublesome to maintain and liable to break in heavy seas or storms, so that though the lightship's crews were often well placed to observe ships in distress, they could not always alert lifeboats on shore. After a series of shipwrecks, an experiment was conducted whereby a nine mile undersea cable was run from a fixed or 'sunk' lightship in the Thames Estuary to the Post Office at Walton-on-Naze in Essex. This was intended to commence in 1884, but was plagued by delays and eventually the trial was abandoned as the cable repeatedly broke. Cables run to light ships were under a constant strain whilst the lightships were at anchor and frequently came into contact with the vessel's hawsers. Trinity house had also investigated Preece's Post Office induction system, but also proved this to be unsatisfactory due to the length of cables required.

Desperate to find some method of reliable communication with their offshore establishments The Elder Brethren of Trinity House, who ran the English Lightship Service, readily accepted Marconi's offer of a trial. They offered Marconi the choice of three lightships, the *Gull*, the *South Goodwin* and the *East Goodwin*. Better still they agreed to pay all expenses.

Marconi promptly chose the *East Goodwin* lightship at it provided the maximum distance for the tests, 12 miles between the ship and a wireless station that was to be installed in the South Foreland lighthouse, some three miles up the Kent coast from Dover.

Behind Marconi's demonstration for Trinity House lay a more ambitious plan. The apparatus he installed at the lighthouse was more powerful than necessary for communicating with the *East Goodwin* lightship lying just over 12 miles off shore. His target was in fact more than twice that, enough he hoped to achieve the first international wireless message across the Channel between England and France. On a clear day you could read the time on the Calais town hall clock from the top of the South Foreland lighthouse, so the distance did not concern him.

The trials stretched over the Christmas holiday with George Kemp on board. Kemp's diary records that he was very cold, wet, miserable, and managed very little sleep. He complained over the radio of pains spreading down his neck, shoulders and spine and was somewhat annoyed about having to spend Christmas at sea. At one point Kemp even pleaded with Marconi to be removed from the lightship, in Morse code, over the ship to shore wireless link. He signalled to Marconi that he urgently needed 'fresh meat, vegetables, bread and bacon'. Marconi had apparently forgotten that when Kemp came on board on 19th December he only carried provisions for one week. After twelve days he found himself having to 'to beg, borrow or steal from the lightshipmen.'

However, despite the atrocious storms, severe illness and being on quarter rations Kemp still managed to repair and maintain the station, train the operators and keep the two stations in regular communication until 9th January 1899, when Kemp finally managed to leave.

During this time Kemp had twice reported ships drifting on to the sands and had exchanged numerous Christmas messages for the wives and families of the crew. Kemp had also taught the light ship crew how to operate and maintain the equipment in less than two days and turned them into reasonable Morse code operators. Overall, as noted and underlined in his diary, Kemp had spent a miserable 22 days on board between 19th December 1898 and 9th January 1899, including the entire Christmas period in the very worst of weathers.

Marconi's trials between the *East Goodwin* Lightship and the South Foreland lighthouse over the Christmas of 1898 had been successful in every way. They had thoroughly demonstrated that his wireless telegraphy communication system was both reliable and robust even under the harshest maritime conditions. The equipment and the operators had also handled emergency situations with ease. It was these events that hit the headlines, but the communication system had also constantly passed general service messages between the ship and the lighthouse together with a large number of personal messages to and from the crew, who led an extremely isolated life. Through it all Marconi and Kemp had tried their best to impress Trinity House with the new wireless system, but as 1899 dawned, inexplicably the Company again failed to get any offer of a contract.

Marconi always understood the value of publicity, especially publicity that surrounded a major step forward, even if that step wasn't in reality a great technical leap. He now set his sights on crossing the English Channel with wireless, not a huge risk as his equipment had long since passed the 22 mile range needed, but none the less a useful milestone and a potentially huge publicity coup. Preece and the Post Office had already looked at the possibility from Dover, but it was Marconi who had his sights set on international wireless communications, an area outside of Post Office control.

Marconi spent time studying the north coast of France and eventually selected a site at the village of Wimereux, site of an abandoned Napoleonic fort, and just three miles from Boulogne where his parents Annie Jameson and Giuseppe Marconi had been married thirty five years before.

On Monday 20th March two of Marconi's assistants brought the wireless equipment over from London by launch, and by 26th of March 1899 the French station was complete.

It was hoped that the much longer aerial consisted of seven, one millimetre strands of copper wire, well insulated and hung from the sprit of a mast 157 feet high at the French station would provide the extra range required. The only concern was that the mast stood in the sand at sea level, with no additional cliff height to aid transmission. The aerial feeder ran across the road, connected to the transmitter and receiver installed in the Chalet l'Artois, on the Boulevard Alfred Thiriez.

The Wimereux transmitter was the standard configuration for Marconi. A spark ball of 2.5 cm diameter powered by a Ruhmkorff coil gave a 25 cm spark. On the station photograph the coil is at the top of the shelf, and the spark gap is in the centre. The receiver included a battery and a relay with coherer tube of 3 mm diameter half-filled with filings of nickel (96%) and silver (4%) under vacuum. The distance of the centre piston was 0.5 mm. The transmitter batteries (cells) were all mounted under the desk.

The attempt to cross the English Channel using wireless over a distance of 32 miles, three times further than the *East Goodwin* Lightship, was ready. With the permission of Trinity House the South Foreland lighthouse station was used along with the same equipment and aerial used for Marconi's earlier trials with the *East Goodwin* lightship. The *Daily Graphic* for 30th March 1899 reported that:

> 'The operations took place, by permission of the Trinity House, in a little room in the front part of the engine-house from which the power is derived for the South Foreland lighthouses. The house is on the top of the cliffs overlooking the Channel.'

On the 27th March 1899 Marconi arrived at Wimereux to meet a Commission appointed by the French government that included representatives of the French Army, Navy and the Telegraph Service. These included Colonel Comte du Pontavile de Heussey, French Military Attaché in England, Captain Ferrie representing the French Government, Captain Fieron, French Naval Attaché in England and monsieur Voisenet, a French telegraph engineer. A special correspondent of *The Times*, Frances M. Merridew was also present.

The French officials had all converged on the tiny town whose squat houses, strung out along the beach, were dwarfed by the towering mast that the Marconi men had raised in the sand. Marconi had taken over one of the small houses on the shore, the Chalet l'Artois. Its new wireless room had flowered wallpaper, a flowered rug and a deal table covered with a flowered cloth on which was sat the strange machine that emitted flashing sparks nearly three-quarters of an inch long. The whole of the wireless apparatus stood on a small table about 3 ft. square. Underneath the table the space was filled with about fifty primary cells while Marconi's usual 10 inch induction coil and exciter occupied the centre of the table.

To the surprise of the dignitaries who gathered to witness the event, the young man in overalls making the final adjustments to the aerial and mast was none other than Marconi himself. When walked in to greet them they were further surprised by how young he looked. At 5 p.m. on the afternoon of 27th March, the demonstration in the packed flower room began.

With Marconi at the transmitting key the first wireless message was tapped out. There was nothing new in this for him except the distance. Months of hard work at the Bournemouth and Alum Bay stations had made wireless communication and building new stations an everyday event in his life. Suddenly, as if he sensed something in the air for him to lend an ear, Marconi signed off as usual with three Morse code Vs as a hand over to the South Foreland station and stopped transmission. The room was silent. Everyone was watching Marconi and their ears seemed to be strained more than his to catch some sound from the receiver. There was a pause but only for a moment, and then briskly the dots and dashes began to click as the tape rolled off the message. There was no panic, no last minute adjustments and no delay. Wireless had almost become ordinary, turn it on, press the key, wait for a reply and then realise that the channel had been crossed in both directions.

'And there it was', said a guest, who later described the historic scene: 'short and commonplace enough, yet vastly important, since it was the first wireless message sent from England to the Continent.'

'And so, without more ado the thing was done. The Frenchmen might stare and chatter as they pleased, here was something come to the world to stay. A pronounced success surely, and everybody said so as messages went back and forth, scores of messages, during the following hours and day, and all correct.'

The incoming message read:

> V [the call letter].
> M [your message perfect].
> VVV

Marconi responded spelling out:

> Same here. 2 CMS [the length of the spark].
> VVV

The first wireless signal ever to cross the English Channel had been successfully picked up by the South Foreland station and then replied to. It was followed by numerous greetings and congratulations in both English and French were transmitted between the little room in South Foreland and the Chalet l'Artois at Wimereux. The English Channel had been crossed.

In the summer of 1899 Marconi was invited to take part in the annual Royal Navy manoeuvres. These were a turning point in Marconi's career both personally and in his relationship with the Royal Navy.

During the extensive sea trials reliable signals sent from HMS *Europa* to HMS *Juno*, achieved when underway at over sixty or more miles profoundly focused Marconi's mind. Alum Bay even signalled to *Juno* over 87 miles. Until that time Marconi was still convinced from all his experiments at Alum Bay, Bristol and especially his last trip to Salisbury Plain that the transmission range for his system was mathematically related to the height of the aerial. The Royal Navy trials finally wrecked all of his various formulae.

In addition what had been considered a freak result, obtained during his trials at La Spezia in Italy where he had communicated with a ship that lay well below the horizon was now confirmed; for some reason wireless didn't travel in straight lines. Hertz and his contemporaries were wrong. Marconi's engineering team now hypothesised that 'Hertzian waves follow around smoothly as the earth curves', but the *Electrician* magazine called this theory 'ridiculous' stating that 'It is an absurdity to suggest that ether waves inherently followed the curvature of the earth.' Whatever the explanation and Physics behind the phenomenon, Marconi had been vindicated. His system worked.

Immediately the Royal Navy trials were completed Marconi was on the campaign trail again. The *Army and Navy Illustrated* magazine for 22nd July 1899 records an account of the first use of wireless in aviation. Marconi demonstrated his system to the Army authorities at Aldershot by installing the transmitter into a captive observation balloon and then using it to send signals to another smaller balloon some miles away. The information was transferred to the ground from the receiver by a wire.

During the first half of 1899 Marconi had increased the reliable range of the wireless apparatus on a boat from eighteen to 72 miles, and boosted the speed of transmission to twenty words a minute. His results at sea during the Royal Navy manoeuvres convinced him that it was time to see America.

Marconi sent a cable on 12th September 1899 to the *New York Herald* which attracted international attention when it published Marconi's acceptance to report on the 1899 America's Cup Race, between the yachts *Shamrock* and the *Columbia* by wireless. Marconi was going to America for the first time. He relished the challenge and the trip as he loved sea travel.

In August, Marconi prepared to sail for America to report on the America's Cup race. He had every reason to be buoyant company prospects. They had never been brighter. He felt that he would solve both the tuning and interference issue. He had set a series of distance records. Lloyd's, Trinity House, and the British Navy were seriously considering purchasing equipment.

He was also delighted and somewhat perplexed by the long ranges (74miles ship to ship, 87 between shore and ship) that his equipment had achieved during the Navy trials. His mood was given another boost when soon after he arrived in America, his new Chelmsford Hall Street factory in Essex received signals from the Chalet l'Artois (23rd September) that had been intended for South Foreland. Chelmsford was 80 miles from Wimereux, 58 of them over land. The next day the Harwich marine wireless station, with standard equipment and 'ordinary' operators also heard Wimereux over 83 miles. According to his current height of the mast against range formula, the Chalet's aerial should have six times higher than it was.

Marconi had only come to America to report a yacht race, by wireless, for a newspaper. However the American Bureau of Equipment was very anxious to have the new equipment tested in order that it might be evaluated for possible American naval use. Marconi's imminent arrival in America, with supporting personnel and apparatus to report the race, was looked upon as an opportunity to conduct these naval tests at small cost and gain an important insight into exactly what the Italian inventor had developed. It would also allow the Americans to fully understand how far behind in the wireless communication race they now were.

Despite poor weather the races eventually started. From the deck of the observation ship which followed the yachts, Marconi transmitted the signals to Bowden who in turn sent them to another Marconi engineer, W.W. Bradfield who was in the *New York Herald* building. From here the reports were transmitted over the land telegraph and via Commercial's cables to the UK. As each update reached the newsroom, the editors' awe intensified. Never in history had an event been tracked or reported in this manner. Over 4,000 words were sent and received in less than five hours between the ship carrying the apparatus and the shore station, from whence the messages were then transmitted over land wires to the papers. The next issue of the *New York Herald* proclaimed: 'Marconi's Wireless Telegraph Triumphs.'

Marconi then started a series of detailed trials for the United States Navy. Unfortunately Marconi had not taken his latest tuneable wireless equipment to America as he had decided it was not needed for the task of reporting on the yacht race. The Navy trials went well, but when presented with simultaneous transmitter stations, the old problems of mutual interference reared its head. Marconi stated that he had not been asked to show his latest equipment. In reality he had already realised that the American Navy and competing American companies were only interested in his ideas and equipment designs. They had no real intention to buy anything.

As soon as the U.S. Navy trials equipment could be packed up, Marconi left America and sailed for England on the American Line vessel SS *St. Paul,* which had been built in Philadelphia, Pennsylvania just four years earlier. The officers of the American shipping line, knowing of the wireless transmitting station at Alum Bay suggested to Marconi that it would be a good idea if news, especially Boer War news, could reach the ship before it reached Southampton. Marconi was always ready for a new experiment or in this case simply a straight forward opportunity for some good publicity, so before he sailed he cabled his head office in London to make it happen.

The ship's owner, the American Line, agreed to allow Marconi to equip the vessel with the wireless equipment he has used for the race and rig an antenna high above deck. Marconi planned to begin transmitting from the ship to his stations at the Needles and Haven Hotels as the liner approached England, to see how far from shore messages could be received.

The Marconi Company's new Managing Director Major Samuel Flood-Page and the previous M.D. Henry Jameson-Davis immediately travelled down to the Isle of Wight and arrived at the Royal Needles Hotel on a Tuesday evening.

Major Samuel Flood-Page

SS *St Paul*

There was no certainty about when the SS *St Paul* would enter the English Channel, or when it would be within wireless range of the Isle of Wight. The Marconi engineers waiting at the Royal Needles Hotel were therefore on tenterhooks. The SS *St. Paul* was *expected* to arrive within transmission distance early on the morning of 15th November 1899. Just in case it arrived earlier an assistant kept watch all night. The engineers had also rigged up a system whereby a bell would wake them if their receiver was called up at night, in the same manner that fishermen attach a bell to their line so that they will know if a fish is biting. But no signal was received.

Flood Page returned to the instrument room at dawn as the sun began to bathe the Needles, a spine of chalk and flint sea-stacks from which the Needles Hotel took its name. The ship did not arrive early in the morning as expected, but at 4.45 p.m. the first signal was received from Alum Bay on board, while the ship was 66 miles away which was replied to immediately. Alum Bay sent:

'Was that you St. Paul?'

'Hurrah! Welcome Home. Where are you?'

Henry Jameson-Davis and Major Samuel Flood-Page were at the hotel waiting the *St Paul's* signal. In a letter to *The Times* Major Flood-Page gave a vivid description of the excitement of the occasion:

'To make assurance doubly sure one of the assistants passed the night in the instrument room, but his night was not disturbed by the ringing of his bell, and we were all left to sleep in peace. Between six and seven a.m. I was down; everything was in order. The Needles resembled pillars of salt as one after the other they were lighted up by the brilliant sunrise. There was a thick haze over the sea, and it would have been possible for the liner to pass the Needles without our catching a sight of her. We chatted away pleasantly with the Haven [the station at the Haven Hotel, in Poole]. Breakfast over, the sun was delicious as we packed on the lawn, but at sea the haze increased to fog; no ordinary signals could have been read from any ship passing the place at which we were.

The idea of failure never entered our minds. So far as we were concerned, we were ready, and we felt complete confidence that the ship would be all right with Mr Marconi himself on board. Yet, as may easily be imagined, we felt in a state of nervous tension. Waiting is ever tedious, but to wait for hours for the first liner that has ever approached these or any other shores with Marconi apparatus on board, and to wait from ten to eleven, when the steamer was expected, on to twelve, to one to two – it was not anxiety, it was certainly not doubt, not lack of confidence, but it was waiting. We sent our signals over and over again, when, in the most natural and ordinary way, our bell rang.

It was 4.45 p.m. 'Is that you *St Paul*?' 'Yes.' 'Where are you?' 'Sixty-six nautical miles away.' Need I confess that delight, joy, satisfaction swept away all nervous tension, and in a few minutes we were transcribing, as if it were our daily occupation, four cablegrams for New York, and many telegrams for many parts of England and France, which had been sent fifty, forty-five, forty miles by 'wireless' to be despatched from the Totland Bay Post Office.'

By then the ship was about 50 miles away. The *St Paul* was the first transatlantic vessel to report its arrival by wireless and then pass on messages from its passengers. These 'wireless-grams' soon became known as 'Marconi-grams', for having been received at the Alum Bay station, they were telephoned through to the Totland Bay Post Office to be immediately dispatched by wire telegraphy to various parts of Britain. By now the rustic Totland Bay Post Office was used to handling unusually heavy loads of telegraph messages, including one giving detailed instructions for the menu at a forthcoming dinner party to be held in London.

On board the *St Paul*, as it steamed towards Southampton, the successful wireless link caused much interest. The operator at the Royal Needles Hotel tapped out a few bits of news, including the latest from South Africa, where the British Army was engaged in the increasingly costly war with the Boers, who had now besieged Ladysmith, Kimberley and Mafeking. With the permission of the ship's commander, Captain Jamison, the on board printers, accustomed to turning out menus and general notices, now produced a small news-sheet souvenir of the occasion under the banner *The Transatlantic Times*, vol. 1, no. 1. It was sold to passengers at $1 a copy, the money to go to the Seamen's Fund.

The 85 subscribers to the world's first 'wireless newspaper' were given the latest news about the Boer War, the paper reporting that Ladysmith, Kimberly and Mafeking were holding out well, there had been no big battle and 15,000 men had been recently landed. The paper also reported the sinking of the U.S Navy cruiser *Charleston*. One news item bristling with British jingoism read:

> 'At Ladysmith no more were killed. Bombardment at Kimberley effected the destruction of ONE TIN POT. It was auctioned for £200. It is felt that the period of anxiety and strain is over, and that our turn has come.'

One of Marconi's engineers, Mr W.W. Bradfield, was credited as 'Editor in Chief', and Henry McClure as 'Managing Editor'. Mr Marconi was recorded as having made the arrangements for the publication. There was even a credit for the treasurer, a Miss J.H. Holman.

The passengers on board ship were extremely impressed with his 'instant news' service, especially as the ship was steaming at 20 knots through thick fog at the time. Marconi was pleased to autograph many of the copies.

Two world firsts were accomplished on this voyage. It was the first occasion in the history of wireless that the arrival of a transatlantic passenger ship was notified to the authorities on land. It was also the first occasion that a newspaper was printed on board a ship using information that was derived from news reports received by wireless.

Marconi had spent years three hectic years testing, trialling and demonstrating his system for the Royal Navy, Army, United States Navy (and Army), Trinity House, the British Royal family and newspapers across the world. He had proved his case. His wireless telegraphy system was no longer a laboratory experiment.

CHAPTER 6

Niton, Tuning
and a
Trans-Atlantic dream

Marconi's Niton station

As the turn of the century drew near, to the outside world it looked as if the advance of wireless communication was unstoppable. Marconi was internationally famous and each day seemed to bring fresh news of another successful trial or a world first. The Company had grown significantly and now had manufacturing facilities in a converted warehouse in Hall Street, Chelmsford, the first factory in the world to build wireless apparatus. Its experimental base was established at the Haven Hotel near Poole and the Company had built other coastal stations including Marconi's first station still operating from the Royal Needles Hotel on the Isle of Wight.

But the construction and maintenance of all these stations together with the considerable sums for wages, travel, materials and equipment being spent by Marconi and his engineers were rapidly draining the Company's coffers. Marconi had started to approach the American market, but in reality the American Marconi subsidiary provided no more than a marker of intention. Despite the advantages of considerable favourable publicity, world-wide fame and the fact that the Company could now offer a viable, if limited, wireless telegraphy communication system, the Company's receipts were insufficient to offset costs and cash flow was becoming critical. Marconi's Company was close to financial collapse.

When the proprietor of the Royal Needles Hotel, William Berkeley asked for another £1 a week in rent, Marconi started looking for somewhere cheaper. In any case by the end of 1899 the ranges that Marconi could now guarantee for his equipment had increased so much that the open sea space available from Alum Bay was inadequate for further useful experiments. He also needed a private facility where his new tuneable equipment and his jiggers could be tested away from prying eyes.

Marconi rapidly surveyed the whole coast of the island, and concluded that St. Catherine's Point, on the southern-most tip of the Isle of Wight would be the site of his new permanent wireless station. The location chosen was an empty cottage at Knowles Farm, near the village of Niton and next to St. Catherine's Lighthouse. The station became known as the Niton wireless station. In autumn 1900 Marconi personally called on the owner of Knowles Farm, Mr. Richard Kirkpatrick, of Windcliffe Niton, to arrange the lease. The Kirkpatrick family were Island bankers and prominent members of the local Congregational Chapel. The equipment at Alum Bay was dismantled between the 22nd and 26th May 1900, packed into vans and moved down the coast to the new station. Marconi and his engineers checked out of the Royal Needles Hotel for the last time. On 7th June the mainmast for the aerial was transported to the new site, but after leaving the main road near Chale, the way was so narrow and the corners so sharp, that a number of walls had to be demolished to allow the vehicle to pass. A new top mast was delivered from George Marvin, a yacht builder at Cowes, and within a week the station was operating on Marconi's newly developed selectively tuned system. Early tests were carried out with trainee wireless operators on HMS *Vernon* in Portsmouth.

Richard Kirkpatrick's daughter, Sylvia Prendergast remembered that her future brother-in-law was working at HMS *Vernon* at the time. He sent her an invitation to lunch by wireless to, 'see if it got through'. She kept the original message tape and some 50 years later she wrote that she gave it to the Science Museum in London.

Marconi took up rooms for himself and two engineers at the nearby Royal Sandrock Hotel. The Royal Sandrock Hotel, a short walk from the new Niton station, was built in 1790 as an inn known as Rock Cottage and it was extended and converted into a hotel in 1812 and again in 1818. It became a fashionable resort after the discovery of a natural spring that was reputed to have medicinal qualities and was visited by royalty in the 1830s. Among the visitors was Princess Victoria, who came with her mother the Duchess of Kent when she was 15, hence the addition of the title *Royal* to the Hotel's name from that date. The Hotel owner's young daughter, Ella Green later recalled how Marconi would take her on his knee and talk to her while sitting on the hotel lawn.

Marconi's system was now reliably producing long ranges, but his detractors still openly maligned Marconi's wireless system for not being able to pass secret messages, or select a single signal if there were many present. In fact they declared that the only thing secret about the system was how it actually worked. They also openly doubted that his new 'tuned' apparatus and his secret jiggers actually worked. Marconi had struggled with the issue of tuning for nearly three years.

But he was close to a final solution. His next patent (Number 5387) was taken out on 21st March 1900. In this the transmitting antenna consisted of two concentric cylinders, the outside one forming the radiator and the inside one being earthed; the receiving antenna was similar. This provided a fair degree of selectivity, but experimental work still proceeded.

Working in secrecy at Niton and the Haven Hotel, Marconi successfully designed his first selectively tuned transmitting equipment. This had always been the Holy Grail of wireless telegraphy development, to create transmitters and receivers that could operate at a single, well defined frequency. To achieve this, Marconi's sophisticated 'four-circuit' design now featured two tuned-circuits at both the transmitting and receiving antennas.

This was it. The problem of full syntony had been solved, and two major worries eliminated at one stroke, for the radiated power was no longer dissipating over a very broad band and consequently reliable ranges suddenly improved considerably. Perhaps even more important, adjacent stations could now conduct their business without interfering with each other.

Niton Lighthouse
The Wireless station and cottage

Niton Wireless station

Niton Wireless station

Niton Wireless station

Niton Wireless station
showing hut at base of aerial mast

Niton Wireless station

Niton Station Wireless Room, c. 1899

Royal Sandrock Hotel, Niton

Royal Sandrock Hotel, Niton

10" Newton Spark coil used at the Niton station

Marconi's new transmitter and receiver designs would allow him to tune his system to a specific wavelength to avoid interference with other signals. It was left to Fleming, with John Fletcher Moulton QC and Major Flood-Page who had joined the main Board of Directors for the Marconi Company as Managing Director in October 1899, to carry out a considerable technical and legal project to strengthen and define Marconi's complete specification in the face of considerable previous work and two other patents already filed by Lodge and Braun.

The team came up with a detailed technical description of the new system used two tuned circuits at the transmitter and two at the receiver. On 26th April 1900, Marconi protected his interests by filing an application for a British patent, number 7,777, on his 'tuned or syntonic and multiplex telegraphy on a single aerial'. The now famous 'four-sevens' patent described the tuned system Marconi had designed. For the first time he announced his jiggers. The American counterpart of this UK patent was No. 763,772, eventually granted in 1904.

An early 'four sevens' patent jigger

Mounted in a square, dovetailed, wooden frame, 20.2cm (9in.) square. Manufactured by the Newton Company, for Marconi's, it has wax-coated windings, ebonite ends and mahogany base 24in (61cm) wide. This was the core of Marconi's first tuned transmitter perfected at the Haven Hotel and tested with the Niton station. The square wooden frame carries two windings. One of these is in series with the Leyden jar and the spark-gap, the balls of which were connected to the terminals of the induction coil. The other winding, consisting of a single turn, was connected between aerial and earth. This arrangement yielded more sustained oscillations and helped greatly to reduce the mutual interference that had been experienced previously. It formed the basis of Marconi's famous Patent no. 7777 in April 1900.

The essence of the 'four-sevens' patent lay in connecting a closed condenser-discharge circuit to an open aerial circuit by means of an oscillation transformer, the jigger. Importantly it covered the use of tuned closed circuits with tuned open circuits in both the transmitter and receiver. It embraced the entire principle of tuning and probably marked the graduation of wireless communication from an intriguing possibility to a viable commercial proposition.

But Marconi's new patent application was always bound to encounter trouble. Oliver Lodge had previously filed a patent for syntonic wireless telegraphy on

10th May 1897 (UK Patent No 11,575, U.S. Patent No. 609,154 in 1898). Karl Ferdinand Braun of Germany had filed a British patent on 26th January 1899 for a very similar design of connecting circuit. (U.S. Patent No 0763345). Furthermore in America, inventor Nikola Tesla on 16th July 1900 had also filed a patent that related to a means for transmitting simultaneously the waves of different frequencies and the means of completing the conjoined recording circuit of the receiving station.

In Germany, Karl Ferdinand Braun had been working independently of Marconi trying to determine why Marconi was finding it difficult to increase the distances over which transmissions could be received. The approach they both used involved increasing the voltage and hence the energy of the spark transmitter discharges and also increasing the size of the aerials. However both men had found that in practice large increases in the spark voltage had resulted in only small increases in the distances that could be achieved.

Braun realised that it was the direct coupling of the aerial that was limiting the range of the Marconi designed transmitter. Braun hastily improvised a test of his loose coupled aerial on 20th September 1898 and filed a British patent on tuning on 26th January 1899, entitled 'Improvements relating to the transmission of electric telegraph signals without connecting wires.'

Marconi had filed his 'four-sevens' tuning patent for transmitters on 26th April 1900, over a year later. But Karl Braun's improvements effectively eliminated the Marconi Company patent monopoly on wireless telegraphy. By 1899 Braun's patents were supported by a commercial syndicate known as *Telebraun*, which was, in turn, taken over by the Siemens-Halske firm in 1901 with the intention of commercialising the system. In 1909 the two men would share the Nobel Prize in Physics but in his acceptance speech Marconi insisted that 'part of my work regarding the utilization of condenser circuits in association with the radiating antennae was carried out simultaneously to that of Prof. Braun, without, however, either of us knowing at the time anything of the contemporary work of the other.'

In typical fashion, despite the earlier patents, it was only Marconi who had set about practically demonstrating the increased range and selectivity which tuned circuits brought to the science of communication by wireless.

Critically Marconi was still living and working on the Company's capital alone, but he decided that his patent was crucial enough to go to America to obtain American patent rights for his new idea. On 2nd June 1900 accompanied by W. Densham, Marconi disembarked from the American Line SS *St Paul* in New York. He had interviews with the Marconi Company's agents, Messrs Moeran and Bottomley, and gave several interviews to the newspaper world, particularly to the *New York Herald.* By the time that Marconi landed back at Liverpool on board the White Star Line's SS *Teutonic* on the 28th June 1900 he had made up his mind that work toward his great transatlantic experiment must start without further delay.

Having reported to the Company Board after his American visit, and secure in the knowledge that the patents had been filed Marconi decided to give a public demonstration of selective tuning between the Haven Hotel and the Niton stations. It was also crucial for Marconi to convince the Company's Board of Directors that further investment was vital. But the art of tuning is not a simple thing to explain to people who have no knowledge of electricity; Marconi had to practically demonstrate to them that he could separate the ripples or as he called them 'isolating lines of communication.'

To make it easier for some of his non-technical board members Marconi likened wireless waves to the ripples on the surface of a pond, and the receiver was to a wood block floating on the pond which represented the *ether* through which the early pioneers thought the waves travelled. A dropped stone was the transmitter. When the ripples reached the block, it bobbed up and down, detecting the presence of the waves.

Marconi now had to explain his 'educated receiver'. That required the wood block to only bob up and down in response to certain ripples and not others. It became even more complicated to explain how the ripples could be separated if two stones, or even ten were thrown into the pond at the same time. Marconi found it was easier to demonstrate it in practice.

Flood-Page was eager to capitalise on the Company's first equipment order from the Royal Navy for 32 sets of equipment order and promote the commercial shipping business of the new Marconi International Marine subsidiary. He prevailed upon Marconi to demonstrate publicly the effects or syntonic tuning. Marconi felt sufficiently comfortable in the results obtained at the Haven Hotel tests to consent.

Newspaper reporters who had long followed Marconi's career were personally called. Flood-Page invited potential corporate customers. Officers of the Royal Navy and Italian and French military and officials at Trinity House and Lloyd's were all contacted.

Approximately one hundred observers assembled at either the Haven Hotel or at the Niton station. At Niton, two operators on two adjacent transmitters at simultaneously transmitted different Morse code messages across the Island and the Solent. The signals, on two distinct bandwidths, had been carefully matched to the identical periodicity and capacity of the two receivers at the Haven Hotel station in Poole.

In the past, the transmissions would have completely interfered with each other and the receivers would have picked up only garbled random Morse letters. But Marconi's new invention tuned one transmitter and one receiver to a narrow band of wavelengths and the other transmitter and receiver to a separate bandwidth.

The two receivers on the mainland immediately and accurately printed the two separate Morse code messages. Applause and commendation from the crowd of witnesses at The Haven Hotel greeted the chatter of the automatic inkers typing out the Morse symbols. It could not have been a more vivid demonstration that two transmitters could operate right next to each other. This moment had been the one eagerly anticipated a year earlier by the U.S. Navy but back then Marconi had been unable or unwilling to deliver. From Marconi's perspective, the wait, despite its costs in possible lost business and adverse publicity, had been worthwhile. His syntonic process now had full patent protection, and the demonstration had been flawless.

Marconi had another surprise. He proposed another, more dramatic experiment. For the first demonstration, each receiver had had its own independent aerial hung from the Haven Hotel mast. Now Marconi placed the two receivers at the hotel literally on top of each other and attached both receivers to the same aerial.

The two operators at Niton, side by side, tapped out two messages on their respective transmitters, one in French and the other in English. The two sets of signals printed out simultaneously and perfectly on the corresponding receiver's tape at the Haven. The message had been received completely separately, despite being connected to a single aerial. In the third experiment, the transmission line between Niton and Poole was crossed obliquely by another signal sent between Portsmouth and Portland. Both signals were received perfectly. Marconi had made his point to his astonished visitors. The next day, the press, in incredulous tones, reported what it had witnessed.

Marconi had successfully 'educated' his coherer and he had effectively demonstrated a new and miraculously improved instrument. More such experiments were performed by Marconi and Kemp between July and September 1900. In late September, for the first time, they succeeded in 'double reception'. Marconi's Niton station sent two messages from two different aerials, and these messages were received separately at the Haven station by two different aerials, each receiving aerial being tuned for each transmitting one. Marconi's new 'syntonic' system even increased the transmitting distance considerably.

In Marconi's own terms, he now had 'syntonic apparatus suitable for commercial purpose'. In any terms, a practical, reliable and tuneable wireless telegraphy system was now a reality. It had taken Marconi, five incredible years of constant struggle. He had solved all the problems and completed his jigsaw puzzle, one he had started just five years before in his father's attic at the Villa Griffone.

One of Marconi's earliest goals had been to challenge the monopoly of the transatlantic cable companies with his wireless system. He wanted to be the first person to send wireless signals across the Atlantic Ocean. In fact he considered it his right.

Marconi pitched his transatlantic experiment to the Company Board as being an aggressive and bold new corporate strategy. If successful, he said, the experiment would eliminate the Company's difficulties once and for all, not only because the Company would control all ship-to-shore communications, but also because it could compete with long distance submarine telegraphy. In essence, if it all went to plan, the Marconi Company could make wireless communication a monopoly.

Initially the Company's Board of Directors objected strenuously. They still argued that increasing the transmitter power by several hundred times, which would be necessary for transatlantic signalling, was impossible. They also argued that despite syntonic tuning and jiggers such an experiment with increased power would interfere with all the other Marconi stations in Britain. Finally, they said that the experiment would be more harmful than helpful, since the Company was now in very difficult financial straits and they doubted the Company could actually cash the cheques Marconi was determined to write.

Their arguments fell on deaf ears. Marconi had made his mind up and unsurprisingly towards the end of July, the master salesman and master manipulator finally obtained the Board's consent. His audacious and risky transatlantic experiment began.

Marconi now believed that he had all the pieces to build and execute his grand plan. He wanted to build two permanent high power transmitting stations that could generate sufficient power to cross the Atlantic. Marconi knew his Atlantic project was fraught with daring and might be a little too much for the public mind to grasp.

It was to be his greatest gamble.

CHAPTER 7

Conclusion

Between 1896 and 1900 Marconi successfully transformed wireless telegraphy from an unstable laboratory experiment into a commercially viable and reliable communication technology. But in 1900 the equipment he could offer for ship to shore communication was still very low powered and relatively crude. Marconi's transmitting apparatus still consisted of a 10 inch induction coil, a contact breaker, Leyden jars, chemical batteries and a telegraphic key, with the jiggers as the only real addition since the first time he ventured out on to Salisbury Plain four years earlier. This equipment could never span an ocean.

But perhaps the most publicised problem to be overcome in his transatlantic adventure was the apparently obvious one. Marconi had already proved that the horizon made no difference to wireless communication. But for the transatlantic bid the receiving apparatus and the transmitting station would be separated by over 2,100 miles, the minimum distance across the Atlantic. At this range there was a mountain of water between the two stations approximately 106 miles high due to the earth's curvature. The best theory at the time suggested that the emitted waves might somehow propagate through the water or be conducted along the surface of the ocean. Given that the mountain of water between there was absolutely no scientific justification to think that such a feat was possible.

This did not deter Marconi. He had every reason to be suspicious of scientists and their unproven theories. Everything he had accomplished so far had been either ridiculed or at best considered of no particular interest by most mainstream scientists. No scientific theory was available to explain how or why his system worked, but work it did, every day over ranges that established scientific theory stated were impossible. Marconi, in an extreme way, represented the triumph of the practical empiricist over the theoretical scientist, a characteristic that had originally endeared him to William Preece, who had been embroiled in such controversies long before Marconi entered the scene.

For Marconi a round world and straight rays just did not fit the evidence. This is where Marconi's lack of formal education proved an asset. The theory didn't matter, only the practical reality. In test after test he had already shown that wireless did in fact work far beyond and below the horizon. He now simply did not care what the scientific community continually told him. Such was the calibre of the man now intent upon transatlantic wireless; the man who was preparing for what he termed, 'the big thing'; wireless communication between the old and new worlds.

Marconi firmly believed that Hertz's electromagnetic waves, for some reason, and with some mechanism yet to be determined would follow the curvature of the Earth. Therefore he reasoned that with stations of sufficient size and power, he should be able to span the Atlantic linking places on either side of the hill of seawater caused by the curvature of the Earth. No one at the time knew that reflections from the ionosphere could greatly extend the range of wireless transmissions, nor the effect that daylight would have on reducing transmission distances. But even the fact that electromagnetic waves would bend did not automatically guarantee transatlantic wireless telegraphy. Success would also depend on the aerial system design and construction, the station's transmitted power, and the sensitivity of the receiver detector.

Marconi built two great stations, one at Poldhu in Cornwall the other at Glace Bay in Nova Scotia. He risked everything on one great experiment, including his entire Company and his reputation. But just months before the experiment was due to commence violent storms destroyed the huge aerial arrays at both his transatlantic stations and wrecked his dreams of exchanging wireless signals across the Atlantic.

It had started off well. While the huge Poldhu Point wireless station in Cornwall was still under construction, at 4.30 p.m. on 23rd January 1901 Marconi succeeded in receiving wireless signals at his new station located on The Lizard near Poldhu. These were sent from the Niton station on the Isle of Wight, 186 miles away, with both stations only using 150 ft aerial masts. This was another world record for long range wireless propagation and for receiving syntonic tuned transmissions.

If the scientists of the day were correct then success would only be possible with masts about 4,300 ft high at each of these sites. It was not just over the horizon, but was some four times the optical range, and twice the distance of any previously recorded transmission. All the other coastal stations in the Channel, including the Haven Hotel, Portsmouth, Portland and Plymouth were picked up during the next few days at The Lizard.

Marconi started to experiment at Poldhu with tuning the system despite the still rudimentary aerial system. Much experimental work of various kinds in connection with the power-control and high-frequency circuits, devised by Marconi himself had been going on at Poldhu since January. By June 1901 the experimental transmissions from Poldhu were being well received at Niton with ease.

Then the storms hit. Marconi was out of time and out of money. In last desperate bid to salvage something at the Poldhu Point station in Cornwall Marconi improvised a temporary aerial array. But the transmitter was only able to reliably generate the shortest of Morse code messages, three short, but very high power sparks that spelt out the letter 'S'.

He was also forced to abandon any chance of two way communication. The Glace Bay station had been destroyed by the storm. Time and finance forced the Italian inventor to go back to the very beginning, and use kites and balloons to carry his aerial wires aloft; just had he had done at his first demonstration on Salisbury Plain five years before.

As noon on Thursday 12th December 1901 approached, Marconi sat in the Cabot Tower in St Johns Newfoundland, waiting with a telephone receiver clamped to his ear. It was an intense hour of expectation. Arranged on the table were delicate instruments ready for a decisive test. Marconi wrote:

> 'Suddenly about half past twelve there sounded the sharp click of the 'tapper' as it struck the coherer showing me that something was coming and I listened intently. Unmistakably three sharp clicks sounded in my ear; but I would not be satisfied without corroboration.'

He handed the telephone earpiece to George Kemp with a quiet: 'Can you hear anything, Mr Kemp?'

Kemp took the headphone. They did not use a Morse recorder or his usual receiver, since the telephone receiver was more sensitive and his receiving system was now not tuned. Through the crash of static he could hear, faintly the unmistakable rhythm of three clicks followed by a pause; then three more and a pause; and so on in constant repetition until, a simple repeated Morse code letter 'S', three dots………dit, dit, dit…….. Then the signals were lost one more in static and the storm. Marconi recalled:

> 'Kemp heard the same thing I did, and I knew then that I had been absolutely right in my anticipation. Electric waves which were being sent out from Poldhu had traversed the Atlantic serenely ignoring the curvature of the earth, which so many doubters considered would be a fatal obstacle. I knew then that the day on which I should be able to send full messages without wires or cables across the Atlantic was not very far away. Distance had been overcome, and further development of the sending and receiving instruments was all that was required.'

Again and again Marconi and Kemp listened to be sure there was no mistake. His other assistant, Percy Paget was called in from the storm. He listened but heard nothing although he was slightly deaf. What Marconi and Kemp heard must have been Poldhu. There was no other wireless station in the world that knew about the pre-arranged signal. It was mid-afternoon. The kite still gyrated wildly in the gale that swept in from the sea. It continually failed to maintain the maximum altitude and its fluctuating height naturally influenced reception. Later that afternoon, at 1:10 p.m., additional signals were received. Around 1:45 p.m. the kite dropped for a short time as the wind eased.

The kite flying team were invited inside to warm themselves and have some hot cocoa that was bubbling on the stove. Marconi seemed very calm and was pleased to receive the quiet congratulations of the men from St. John's on his successful reception. Shortly after 2:00 p.m. the wind strengthened and again the kite was raised for the final time that afternoon. At 2:20 p.m. another set of signals was received.

There were 25 clicks in all. Kemp, who kept his detailed diary, recorded the event as if it were scarcely nothing more than an account of putting his boots on.

```
Marconi tried all the detectors from time to time.
Signals appeared at intervals on a telephone in series,
when using our sensitive tube (coherer) circuit, and,
at times, the dots threatened to appear on the tapper.
```

Marconi was no diarist, but occasionally he scribbled laconic entries in a pocket book. He simply logged the historic event; *'Sigs at 12.30, 1.10, 2.20'*

The Atlantic had been bridged by wireless and history had been made. It was five years to the day since Marconi first stood alongside William Preece in Toynbee hall to demonstrate his new system of telegraphy without wires.

His transatlantic gamble had paid off as Marconi always *knew* it would. It had been a gamble born out of necessity at a time when his Company had reached a commercial deadlock. Marconi's transatlantic experiment was conceived in part as a means to overcome this. Its success was to prove a forceful weapon with which to quieten his enemies, achieve worldwide publicity and an opportunity to increase the capital investment in the Company.

He had also spent a fortune chasing his dream. To the ever practical Marconi, despite his commitment to the ongoing development of the transatlantic wireless service it was clear the most promising field of financial security was still marine installations.

From 1900 to 1910 there was still little profit to be found in the wireless telegraphy business. Despite a boom in wireless telegraphy share prices on the stock market, the Marconi Company had still not seen any tangible commercial profits and had been continually challenged by aggressive competitors in Britain, Germany, and the United States who bypassed or simply flouted Marconi's patents.

It was not until the tragic sinking of the *Titanic* in 1912 that public attention was focused on the importance of ships being equipped with wireless and the necessity for maintaining a 24 hour communications watch. After the loss of the *Titanic* many countries required that ships over 500 tons be equipped with radio. This caused demand for marine equipment to soar. In 1913 the Marconi Company was finally able to declare its first ever dividend to its shareholders. They had waited a long time. Within a decade, over 500 stations had been built, establishing wireless, now almost universally known as radio, communication worldwide.

The transatlantic experiment brought Marconi honours from around the world, great wealth and undying fame. The whole adventure was far more than just another experiment in the early days of wireless, for Marconi showed for the first time that the four corners of the world could be joined without wires. For a brief moment on one fateful November afternoon in 1901, he linked an uncertain kite with a shattered experiment and allowed thought to pass between them, crossing an entire ocean in an instant.

His success also gave him the optimism and courage to persist, and the impetus imparted from the transatlantic success was destined to make radio into the greatest mass medium for entertainment and communication the world has ever seen.

In just five years the Italian engineer had changed the world, a world where communication had been dominated by cable telegraphy, but with no air travel, few cars, and limited telephone communications. He combined the world's scientific knowledge with his own exceptional drive, experimental skills and business acumen to create the face of the radio, broadcasting and electronics industries for the next century.

The Company that Marconi had started on a windswept headland with a small band of pioneering engineers from Alum Bay, eventually employed over 48,000 people worldwide. What they accomplished in the last years of the nineteenth century would affect mankind for the rest of time.

The Age of *Radio* had arrived.

CHAPTER 8

End Words

On 19th July 1937, at his office in Rome, Marconi met his old friend Luigi Solari to discuss the microwave experiments he was planning on board his floating laboratory, the yacht *Elettra*. He told Solari: 'There is a great deal yet to do in this field.....I wish I had the energy I used to have...the energy I no longer have.' He was 63 years old.

His wife was away overnight and the next day was their daughter's seventh birthday. Returning to their apartment in Via Condotti, he cancelled a Royal Academy appointment and went to bed. The last person to see him alive was his physician, Professor Cesare Frugone. 'How is it', Marconi asked, 'that my heart has stopped beating and I am still alive?' He died soon afterwards, at 3.45 a.m. on 20th July 1937.

Throughout the following day, Marconi's body lay in state in the Farnesina Palace in Rome. The day was hot and a crowd numbering in the thousands packed the square in front of the palace and filled the surrounding streets.

At six o'clock that evening, when his funeral began, wireless operators around the globe stood to attention and their transmitters fell silent for two minutes. For possibly the last time in human history, the 'great hush' again prevailed. Guglielmo Marconi was laid to rest. No other individual, in any field, before or since, has been accorded such a tribute or such universal respect. It was a sublime irony that the man hailed as the father of modern communications, the man whose genius gave the world a voice, was honoured by silence. There could be no more appropriate recognition of Marconi's contribution to the story of mankind.

The news of his death was announced to the world in the way that he would surely have liked best; by wireless. Shortly afterwards flags were flying at half-mast in nearly every capital in the world in honour of the great inventor; and tributes were paid in the newspapers of every nation. *The Times* wrote 'When the early 20th century comes to be surveyed by historians yet unborn, Guglielmo Marconi may be regarded as the supremely significant character of our epoch'.

Of course without his work, someone else would have eventually broken the silence. Had Marconi not seized the opportunity and commercialised wireless when he did, then inevitably somebody else would have done it. Marconi was always surprised by the great rapidity with which competitors appeared and embraced any new technology and the speed with which many of them became commercially successful; some of them embarrassingly so.

Marconi left behind a legion of detractors who pointed out that other scientists and inventors had sent some form of wireless signals before Marconi got his patent. But it does not really matter. What Marconi undoubtedly did was to invent what became an entirely new industry. In his hands an obscure and to most people unintelligible branch of physics and electrical engineering became a simple consumer product. Throughout history there are very few inventions or technologies that actually changed the world. Marconi's is one of those.

It is likely that without Marconi's determined and driven stewardship there would not have been the explosion in communications technology as we know it.

Quite simply, in just five years, Marconi's vision transformed our world and the silence was broken by the youthful dreams of one man. Marconi left behind him a world that had come to regard wireless as a commodity, not a miracle or even *a kind of magic*.

There was a time at the end of the nineteenth century when the skies were 'painted with unnumbered sparks', but those first, awe-inspiring spark and discharge machines are now all long gone. There are no more crashing, cracking sparks; no more electric flames leaping across wide-open gaps, no more induction coils and coherers. The long Hertzian waves upon whose crests Marconi built the art and science of wireless lost much of their romance, as short radio waves became the new carrier and large industries became the new pioneers.

The battle for wireless changed. The name even changed. Building on Marconi's ideas, softly glowing radio vacuum tubes silently broadcast a rhapsody of words, ideas and thoughts across the world without as much as a blink. Nation spoke unto nation. But perhaps the science that was created in a vacuum, that enabled silence to be dispelled, made wireless even more mysterious. It was just a newer kind of magic.

Marconi's dream started on the Isle of Wight, at the wireless station above Alum Bay; that was radio's first home. From that windswept bay to the storm-lashed Cornish coast he fought for his dream. It was there that he gambled everything as looking west from both places there is nothing but cold and inhospitable ocean for over two thousand miles.

On the high cliffs, above the foaming sea at Alum Bay and at Poldhu Point there stand monuments to a young Italian inventor called Guglielmo Marconi and his experiments that are now well over a century old. What Marconi achieved there changed the world forever and made it a far smaller place to live in.

Easily passed by so many without a second glance, if you ever have the time to visit take a moment to read the plaques and wonder at the story they tell.

Amazing things happened at these places.

Guglielmo Marconi, c. 1901

APPENDIX ONE

The Isle of Wight and South Coast Sites

Today

Despite the fact that the invention of wireless revolutionised the world, the places where the development and technical breakthroughs actually took place are little known. The sites where wireless was born and subsequently grew into a viable communication system must be considered as parts of this country's industrial archaeology and heritage. Although they often lie off the beaten track, most can be easily found. Part of the idea behind this book was to visit every site today and record and photographs what remains at each one.

To the locals the names are often familiar, but I have found in my travels that few know of the sites' history. Now with this section you can visit the places where Marconi worked on the Isle of Wight and South Coast and built the science and industry of wireless communication.

Alum Bay

The new Royal Needles Hotel
Converted from original Hotel stables

Alum Bay lies on the west tip of the Isle of Wight, and has been a favourite holiday and tourist centre for over 200 years. In the latter part of the 18th century the first tourists started to arrive at Alum Bay and during the early part of the 19th century it became an essential place to visit during any Island holiday. Visitors were catered for by the Royal Needles Hotel on top of the cliffs and a pier brought tourists from the mainland that marvelled at the coloured sands, and used them to fill shaped clear glass bottles to take away as souvenirs. The creation of ornaments using the 21 different coloured sands that are layered in vials and jars and also used for pictures soon became a popular craft in Victorian times known as *marmotinto*.

Having chosen Osborne House to be her new family retreat, Queen Victoria was the prime mover in the rapid growth of this former backwater, local artisans benefitted from the influx of wealthy visitors, and a number of craftsmen sold their fixed sand pictures and unfixed sand jars featuring views of the Island as unique keepsakes of the Isle of Wight. Some of the early layered sand pictures and filled glass shapes are now highly collectable.

The easiest approach to Alum Bay in calm weather was by sea, to which end in 1869 the Alum Bay Pier Act was passed by Parliament. The first pier was a simple wooden structure that could not keep up with its unforeseen popularity with constant steamers calling, and so a second Pier Act was passed in 1887, with a replacement 370-foot long pier opening in August 1887. The pier was 'paddle shaped' so that the pier head was twice as wide as the promenade leading to it. On the pier was a cafe, gift shop and a small restaurant, where a glass of water cost 1d. The pier was often very busy, no more so than on the 23rd July 1909, when the 9,060 ton German liner *Derfflinger* ran into the shingle bank off the Needles, and two and a half miles from Alum Bay. She was trapped for two days before the tugs nearby were able to pull her free, all of which was witnessed by people who packed onto the pier.

Paddle-steamers regularly called at Alum Bay, including the *Princess Beatrice*, *Prince of Wales* and *Lorna Doone*. Even Islanders enjoyed travelling to Alum Bay on the paddle-steamers and day trips were available there from Yarmouth. Steamers on the London and South Western Railway frequently called there until the outbreak of the Great War in 1914.

The pier was run by Alfred Isaacs, a descendant of one of the first lighthouse keepers, who with his family owned land nearby. Isaacs was to be a great help to Marconi over the year he spent at Alum Bay helping organise local labour and transport. There were a range of bathing machines and small boats available for visitors to hire, as well as a café on the pier and a golf course constructed nearby. From the top of the cliff to the beach below, a path comprising 232 steps was constructed.

The Needles Hotel, where Marconi established the world's first permanent wireless station was built in 1874 to accommodate the increasing number of visitors to the coloured sands and to the now vanished Alum Bay pier. Twenty years later it was renamed The Royal Needles Hotel. A serious fire fanned by very high winds in the early hours of Sunday 20th January 1910 caused extensive damage to the hotel and there was no mains water supply to fight the fire. There were rumours on the Island that the owner, William Berkley had deliberately set fire to the hotel for an insurance claim as he had recently retired to Totland. There is an Island legend that all the staff were sent home for the weekend and the Hotel's silver was removed, but nothing was ever proved. The ruins of the front facade were still standing as late as 1928, but eventually it had to be demolished and the site was lost forever.

With the Great War and the following depression, the tourist trade at Alum Bay went into decline. The stables of the Royal Needles Hotel were converted into a restaurant by Frank Cotton. In 1924 he sold the Royal Needles Hotel's land and restaurant to the Needles Hotel Company, rivals to the Isaacs family, who still owned the pier, much of the rest of the Alum Bay area and The Hut café.

The Royal Needles Hotel, burnt out shell. c.1915

The Royal Needles Hotel, c1928

After the First World War, the Alum Bay pier was virtually abandoned. The last steamer to call at it was Red Funnel's *Queen* in 1920. By 1925 the pier was declared unsafe and closed. In the winter of 1927 the pier suffered severe storm damage, and broke in two. Only the shore-half remained, and was still in use when Alum Bay was used as a military practice area in World War II. The military removed the majority of the remains during the Second World War, but the remnants were visible well into the 1960s, although today no trace of the pier remains, and Alum Bay has just a small landing stage. While the Isaacs family were recovering from this financial catastrophe, the new hotel Company outbid the Isaacs and bought the beach and cliffs south of the pier, although the Isaacs owned the well that was Alum Bay's water supply.

Alum Bay – c. 1924

Royal Needles Hotel, c.1908

The renamed stables block for the Royal Needles Hotel is in the centre, Headon Hall behind it and the remains of the original Hotel and wireless station can be seen to the left.

Photographs show that the Hotel was situated adjacent to the chair lift and the aerial was situated in front of the Hotel to the seaward side, right on the cliff edge. Extensive alterations and collapses over the past 100 years to the whole headland mean that the exact location cannot be found. Much of the actual site of the hotel has long since passed over the cliff.

During the two World Wars the area was heavily militarised and access to visitors was barred. The Needles headland was used as a military training area, and many of the buildings were damaged. However tourism began again in 1946 and in 1950, on the death of Sam Isaacs, the Isaacs family sold their land to the Needles Hotel Company, with Jim Isaacs as its Managing Director until his death in 1968.

By the 1950s Alum Bay had fallen into serious decline, but was revived when hundreds of scientists and engineers moved to the West Highdown in 1956 to begin testing rocket engines for the British Inter-Continental Ballistic Missile program. During the peak of activity in the early 1960s some 240 people worked at the complex and successfully built, assembled and tested the engines for Black Knight and Black Arrow rocket between 1956 and 1971. The rockets were built in nearby East Cowes and were later used to launch the Prospero X-3 satellite.

To assemble and test each rocket before shipment to the Australian launch site at Woomera, Saunders Roe selected the former artillery battery on the Needles Headland as it offered a secure location with underground accommodation.

In 1955 the site was leased from the Ministry of War and from April 1956, the engines were assembled, tethered and fired, with different levels of fuel to measure the thrust, flight control systems and the consumption of the fuel. A complex of specialised buildings was constructed over the 'New Battery' built in 1893 to take three 9.2-inch breech loading guns and altered and enlarged in 1900-02, 1911-14 and 1939-45. It comprises a concrete redoubt curved gun emplacements with semi-underground magazines between. For the rocket engine tests the underground control and instrumentation rooms were converted from the old battery magazines.

There were 2,200 square feet of control rooms and underground stores, 4,260 square feet of laboratories and offices and 3,080 square feet of workshops and smaller machine shops. The dining rooms catered for 80 people at a time. In all there was space enough for the 240 people who worked there at the peak of operations in the early 1960s. The site closed in 1974 and was sold to the National trust in 1975 but the large concrete engine testing structures and many of the associated buildings remain to this day.

From the early 1970s major renovation work took place at Alum Bay and the last traces of the Royal Needles Hotel were lost. In 1971 construction began on a chairlift to take visitors from the cliff top to the beach. Opened in April 1973, it travels 250 metres across and 51 metres down to the site of the pier. It is capable of carrying 2,000 passengers an hour, 1,000 each way. As part of the refurbishment, the ageing collection of huts that clustered around the top of the chine was turned into a modern tourist attraction with a bar, café, souvenir shops, fast food outlets, glass blowing studio, sweet factory and amusements. Other attractions include the coloured sand shop, crazy golf and even a growing collection of children's rides. The area also includes a replica invasion beacon and the Marconi memorial. The Needles Park at Alum Bay is visited by about 450,000 tourists each year and has a full time staff of about 40 and employs about 100 seasonal staff. The present Needles Hotel (which actually has no rooms) and the 'Marconi' bar nestle amongst the shops and stalls. The current restaurant kitchens were actually the old stable block for the original hotel.

The best way to see the remarkable phenomenon of the coloured sands is to use the chair lift which runs between the cliff top and the beach, but aspiring sand crafters are now banned from risking their lives climbing the cliffs to collect the coloured sands available in the bay. This also prevents excessive damage to the environment, but the sand kiosks in the shopping complex above now supply their needs.

On Saturday 14th July, 2012 the Olympic torch arrived at the Needles Pleasure Park and was taken on the chairlift down to Alum Bay before returning to the top of the cliff.

The Alum Bay wireless station and the pioneering work that Marconi undertook there are commemorated by a large granite column. In the past 15 years this has been moved to a new site closer to the sea and the original Hotel site and away from the amusement arcades. The hordes of summer holiday makers heading for the chair lift, ice cream stands, shops and even the famous coloured sands usually pass it by, but the monument's four plaques tell their own unique story:

'This stone marks the site of the Needles Wireless
Telegraph Station where Guglielmo Marconi and his
British collaborators carried out from 6th December

1897 to 26th May 1900 a series of experiments which constituted some of the more important phases of their earlier pioneer work in the development of Wireless Communication of all kinds.

The Needles Wireless Telegraph Station exchanged radio messages first with a tug in Alum Bay then with Bournemouth 14 miles distant next with Poole 18 miles away, later with ships 40 miles seaward. These wonders attracted world wide attention and famous scientists from many countries came (1898-1900) to see the new wireless telegraphy in experimental operation.

On 15th November 1899 information for the first newspaper ever produced at sea the Transatlantic Times was transmitted from this station by wireless telegraphy and printed on the U.S. liner 'St Paul' when 36 miles distant. On 3rd June 1898 Lord Kelvin sent from the Needles Wireless Telegraph station the first radio telegraph for which payment was made.

Marconi described the Needles station as the world's first permanent wireless station. It was erected under his personal supervision by his assistant George Kemp for Marconi's Wireless Telegraph Co. Ltd. and was completed on 5th December 1897. Other radio technicists of this Company who pioneered here were P.W. Paget, A. Gray, C.E. Rickard, W. Densham, F.S. Stacey, P.J. Woodward, C.H. Taylor. The station was dismantled in June 1900.'

Needles Monument, Alum Bay

Alum Bay Monument Plaque

Site of the Royal Needles Hotel, 2012
The hotel was located roughly where the motor homes are parked

Alum Bay Aerial View 2000

Alum Bay Aerial View 2000

Alum Bay Site Chair Lift

Alum Bay, The Restaurant and
'Marconi's' Bar
The Left hand side is the original
stable block

Alum Bay – Marconi's Bar, 2012.
The left hand building was the stable block for the original Royal Needles Hotel

Totland Post Office

The popular story is that in April 1897 Marconi walked into the Totland Bay Post Office and asked the Postmaster, Mr J. B. Garlick if he could assist him with his experiments in radio transmission a mile up the road at the Alum Bay wireless station. In fact both Marconi and Kemp were proficient Morse operators but the Post Office was for a short time asked to carry an extra load of wire-based telegrams as Marconi's experiments grew. A young lad called Sam Isaacs was employed to run messages to and from the Marconi station and up to the old coastguard cottages and on the Headland. Most transport would have been by a horse drawn Hackney Carriage. The charges, still displayed in Totland's Post Office (the front is now a public bar), are for one horse, 3 shillings 5d for one hour, (5 shillings for 2 horses) and every additional 15 minutes cost 9d. There was also a distance charge, one mile was 1s.6d and each extra mile was one shilling.

Built around 1870, originally the Post Office was the Totland Bay Supply Stores which sold everything from groceries to coal, beer to bread and it later became the Post Office. By 1902 Totland had started to expand and the post office was relocated 300 yards away. The Supply Stores were renamed Jordan & Stanley and ran as such until after the Second World War. In 1948 an application was made to turn the shop into a public house, the Broadway Inn was born in 1951.

The Totland Post Office was closed in 2006, but it re-opened actually inside the Broadway Inn in 2009, following campaigns from locals to bring a village post office back to the area. Outside the post office is a long thin red board.

IN APRIL 1897
SIGNOR MARCONI
WALKED INTO
TOTLAND BAY
POST OFFICE
AND ASKED MT GARLICK, THE
POSTMASTER, FOR ASSISTANCE
IN HIS EXPERIMENTS
IN TRANSMITTING WIRELESS
MESSAGES. HELPED BY MR
GARLICK HE SET UP HIS
TRANSMITTING STATION
AT THE ROYAL NEEDLES
HOTEL. ERECTING A HUGE
MAST WHICH HAD COME FROM
THE ROYAL YACHT *BRITANNIA*.
MESSAGES WERE SENT AT
FIRST A DISTANCE OF ONE
MILE TO TOTLAND, THEN
EIGHTEEN MILES OUT TO SEA
TO THE STEAMER *MAYFLOWER*.
FINALLY THE U.S. *ST PAUL*
PICKED UP A MESSAGE
THIRTY-SIX MILES OFF
THE NEEDLES

A MEMORIAL TO THIS EVENT
STANDS AT ALUM BAY

Marconi Board at Totland Post office, 2012

Totland Post Office, 2012
The Post Office was in the Left-hand
building

Inside the Totland 'Post Office'
Looking toward site of Post Office, 2012

Fort Victoria

Fort Victoria was a single tier battery with defensible barracks built at a cost of £38,000 between 1852 and 1853 to defend the Solent. First garrisoned in July 1855, it is situated west of Yarmouth, Isle of Wight a few miles from Totland and Alum Bay. In 1891 the 22nd company of Royal Engineers were based there from which coastal minefields were laid to defend the Needles passage. The Fort was a base for experimental searchlights mounted in the hillside behind the Fort from 1898. It was later used as a submarine mining centre and training area for military purposes. The last soldiers left the site in 1962 and the larger barrack blocks were demolished in 1969. Two remaining sides of the sea-facing casemates were included in the Country Park and today form a small tourist attraction and shops.

Fort Victoria Pier was built in the 1850s to service various military installations in the West Wight and was used as such until 1962. It had a small tramway on it and a crane at its head, but today it is derelict and closed off. During the first months of the Alum Bay station operation Marconi and Kemp established a receiver station at the Fort to calibrate and monitor the Alum Bay equipment. It is highly likely that Marconi made use of the pier (as he did in Alum Bay, Bournemouth and Swanage for his initial experiments) as its position over water gave his aerial system a much better ground plane for operation. Although Totland pier was closer to Alum Bay it sits beneath a high headland and Marconi was not yet confident that his wireless waves could pass through or over intervening land masses.

Fort Victoria

Fort Victoria Pier

Ladywood Cottage

The equipment for the wireless station established in the grounds of Osborne House for Queen Victoria was installed in Ladywood Cottage, roughly half-way between the main House and Osborne Bay, well to the north of Swiss Cottage. There is no plaque or monument to this Royal connection with early wireless, and the experiment has no mention in any current guide book or tour. Ladywood Cottage was demolished in 1912 and no traces are visible as its exact location was lost when the Osborne House golf course was constructed in 1972. The cottage was located on what is now the Northern edge of the 7th hole fairway, at the top of the hill that still offers an excellent view to the sea. There are to my knowledge no known photographs of Marconi's activities at Osborne and the site of Ladywood Cottage is unfortunately not open to the public, unless you decide to play the nine hole golf course.

Osborne House Golf Club

7th Fairway, site of Ladywood cottage

The Niton Station

Niton village on the Isle of Wight is split into two halves by a break in the inner cliff large enough to house the main road that runs through Niton. Upper Niton lies in a hollow and is set around a crossroads. The lower part of the village, below the inner cliff is often known as Niton Undercliffe, and was a small fishing hamlet until the 19th Century. This part of Niton flourished in Victorian times due to the popularity of Ventnor as a health resort and many mansions and holiday cottages were built here. The road on the Undercliffe continues east from Niton towards Ventnor. The Undercliffe at Niton includes the most southerly point of the Isle of Wight, St. Catherine's Point, and St. Catherine's Lighthouse.

The site of the original Marconi Niton wireless station lies well outside the village, next to the lighthouse on St. Catherine's Point. The present three-tier octagonal lighthouse was completed in 1840 following the shipwreck of the *Clarendon* in Chale Bay. However, the tower was later lowered as its light was often lost above the mist. The newer Niton radio station shadows the headland from the hills above, but has nothing to do with Marconi's early work at Niton.

In 1270, Richard Knol was bailiff of Niton (one of the King's officers charged with local administrative authority). It is thought that Knowles Farm was named for him. It forms part of the estate situated at the western edge of the Undercliff, which stretches from Gore Cliff in the north down to the shore. The estate has been owned by various eminent families. These include the Worsleys, many of whom were baronets and MPs on the Isle of Wight, and the Meux family of London brewers.

Niton Lighthouse and Knowles Farm

Knowles Farm

Knowles Farm

National Trust Plaque

In autumn 1900 Marconi personally called on the owner of Knowles Farm, Mr. Richard Kirkpatrick, of Windcliffe Niton, to arrange the lease. The Kirkpatrick family were Island bankers and prominent members of the local Congregational Chapel.

To get to the site, walk down the road to the lighthouse, but fork right on to a rough track by the National Trust sign. The station was located in the group of cottages at the end of this track known as Knowle's Farm. They are now owned by the National Trust, but are private property.

The radio mast that Marconi used to make his pioneering transmissions from Knowles Farm was situated in the field to the south of the end farmhouse, about 30m from the nearest building toward the sea.

The mast was removed in the 1920s, when a farmer chopped up the now redundant mast to make ladders and the last lump was apparently taken by a local sculptor. The small concrete base on which it stood, about a metre square can still be found part buried under the long grass in the field to the south of the farmhouse. (The author is pictured on the inside back cover taking a rest on the aerial base). The field is entered by a gate on the left immediately before the cottages, and the aerial mount lies in the field directly behind the end cottage, complete with its interlocking spikes and several large bolts. The zinc plate ground-plane and some of the aerial guy bolts are still thought to be buried around the mast.

The cottage furthest from the lighthouse is where Marconi built his station and it is shown on early maps back to 1793. It now bears a small memorial tablet set into its wall.

‘This is to commemorate that Marconi
set up a wireless experimental
station here in 1900 AD’

Niton Station Plaque

148

It has been said that Marconi lived in the farmhouse and used the cottage next door for his pioneering radio experiments. Indeed this middle cottage is now called Marconi Cottage (renamed in the 1970s) but the name is a slight misnomer as in 1898 it was a newly built chicken run and unlikely to be useful or suitable for wireless equipment. It seems much more likely that Marconi installed his equipment and worked in the right hand farm houses back room that is directly in line with the aerial mast base. Marconi himself stayed at Sandrock House Hotel a short walk up the lane, but his engineers would most likely have slept upstairs at the farm house although in the earliest days due to battery and spark fumes this would not have been the best of environments.

The 1899 photographic evidence of the station when compared to the room today supports this. In 2012 I visited the house with the kind permission of the owners. In what is now the dining room, the book case to the left of the more recent chimney breast can be seen in both photographs. It was reputedly built by Marconi to store his books and notes. New modern windows have been installed in place of what appears in the original photograph to be two farm house openings. In my opinion this is undoubtedly the oldest surviving wireless station in the world. The cottages are private property and should be respected as such, but the plaque and aerial site can be easily seen without disturbing the owners.

Niton Station, from seaward side, 2011

Niton Station, 2011

Niton Station, 2012

Niton Station aerial mast base, 2011

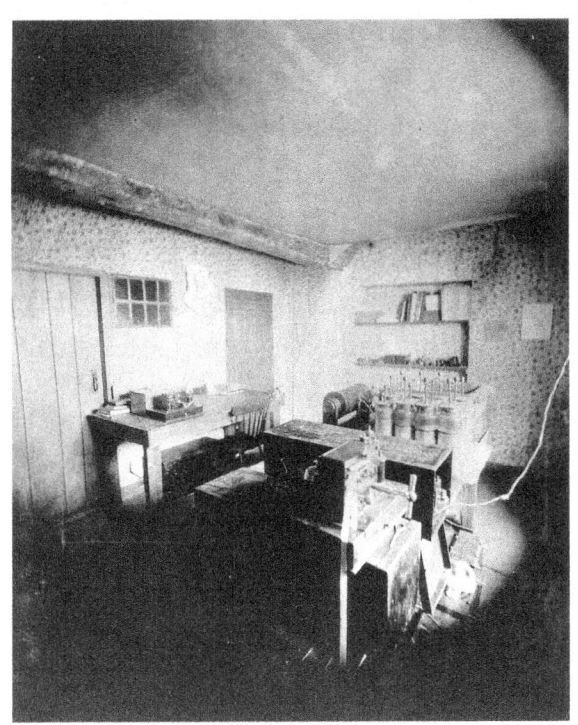

Niton Wireless Room, 1899, 2012 and then and now....

Niton Radio Station GNI

As the orders for ship wireless equipment started to increase, Niton soon became an active Marconi shore station, handling passing ship wireless traffic in the Solent.

Marconi's Niton station was taken over by the Post Office on 29th September 1909 and four years later, as part of a major reorganisation, land at Niton Undercliffe, about four miles from Ventnor, was leased from Lloyd's at an annual rent of £5.

On the new site a Lloyd's signal and wireless telegraphy station using Marconi equipment was built, along with various houses which belonged to the Coastguard, and the station building that was later to become Niton Radio.

Lloyd's Signal station, Niton, c. 1910
The Lloyds Signalling Station was at the top of Castlehaven Lane.
It was demolished in the nineteen eighties

Lloyd's Signal station, Niton
Shown to the left hand side of the photograph, St Catherine's lighthouse is in the
middle and Marconi's Niton station to the extreme right.

After Marconi obtained the patents for the Bellini-Tosi Direction Finding (D/F) system, successful tests were carried out at Niton in 1921 which resulted in all coast stations being equipped.

The lease on the land at Niton Undercliffe finally expired in June 1951 and the Post Office bought the site for £7,000. Equipment was changed during the 1950s in line with Post Office upgrading plans and in 1959 Niton was one of the first stations to be fitted with VHF radio. Unlike other stations, however, the equipment could not be sited at the station itself. The main station was on comparatively low land in the shelter of a cliff face, which was unsuitable for a VHF service, which was required to cover not only the open sea but the approaches to Southampton. A reasonably high location was needed and an old Air Ministry site at St. Boniface

Down about 800 feet above sea level and four miles from Niton provided the answer. Suitable buildings were already there and the VHF equipment was located in an old bunker. A 180-foot mast was erected for the aerials, and control of the equipment from Niton was by landline. So after 60 years of sterling service the Marconi station at Niton finally closed, moving up the hill.

Niton Radio (call sign GNI) was maintained as a coastal radio station well known to yacht masters, including being featured in a British Telecom International information film until it finally closed, along with the rest of the coastal radio network, in 1997. During May 1997, the United Kingdom coast radio stations broadcast the following information at their weather broadcast times:

The following MF and VHF channels will close at midnight on 31st May 1997:

NITON RADIO - MF Channel Uniform
SHETLAND RADIO - VHF Ch.27
NITON RADIO - VHF Ch.85
CLYDE RADIO - VHF Ch.26

At the same time, NITON RADIO/GNI will cease all WT services.
HASTINGS RADIO - VHF Ch.66 will cease on 16th June 1997.

After 97 years, there ceased to be a manned ship-to-shore radio communications station on the Isle of Wight.

The Royal Sandrock Hotel, Niton

While working at the Niton wireless station Marconi and his engineers stayed at the Royal Sandrock Hotel. After a period of decline during the mid nineteenth century, the Hotel had evidently recovered and by April 1891 the annual census showed that Rudyard Kipling, already a celebrated author was a resident there. He was visiting with his American friend and fellow author and collaborator, Wolcott Balestier.

After Marconi left, the popularity of the hotel gradually declined especially after a large rock fall permanently closed the road to the west in July 1928. The Royal Sandrock Hotel was fatally damaged by a fire on 1st October 1984. It was then demolished and the Undercliffe lost its favourite place of entertainment. It has now been replaced by three modern red brick houses.

Royal Sandrock Hotel – Fire Damage, October 1984

The Site of the Sandrock Hotel, 2012

Madeira House Hotel, Bournemouth

The Madeira House Hotel wireless station started operation on 20th January 1898 until the station was moved along the coast to the Haven Hotel, in Poole in October 1898. The Hotel continued to operate until the Second World War when it was used by the Royal Air Force, and then fell into disuse. In 1947 it was purchased as a Convalescent facility for the Miners of South Wales through funding from the Miners' Welfare Committee.

The convalescence on offer at the now renamed Court Royal Hotel was available to mineworkers who had suffered an injury in the course of their employment in the coal industry. Following their recovery from injury, men would be allowed a two week period of convalescence in Bournemouth before recommencing work. The accommodation at Court Royal originally consisted of three dormitories housing a total of 60 people. Each dormitory offered only a single toilet and it has been suggested that the seat was 'never cold in those days'. The regime within Court Royal was much stricter then than it is now as everyone staying there had to be in by 9.30 at night and in bed by 10.00pm.

In 1972, the management committee decided on a programme of renovation for Court Royal and the dormitory accommodation was phased out in favour of en-suite rooms. Around this time, partly due to the improvements in working conditions within the Industry, fewer injured workers were using the facility and it began providing more convalescent holidays for retired mineworkers.

During the 1984/85 Miners' Strike, the future of Court Royal was in some doubt as the strike led inevitably to a lack of funds for the Home. As a consequence of this, it was decided to allow retired mineworkers and their wives to purchase holidays at Court Royal. In 1985, wives of miners were allowed to use the Court Royal between April and September, but they would have to pay. The men could still use the facility from January to March, and October to December for 1 or 2 weeks free of charge. This move was revolutionary at the time but it has allowed the facility to develop into a fine 3 star comparable hotel.

The Hotel enjoys unrivalled views of the sea front and pier at Bournemouth, on the sunny south coast. The Court Royal Hotel is now open all the year around to former mine workers, and wives, and indeed mine workers' widows, and thousands of miners have stayed there since 1947.

Madeira House, 2012

Madeira House, 2012
Looking from the pier

When Marconi died in July 1937, *The Bournemouth Times and Directory* reported: 'Bournemouth may take special pride in the fact that it was here that senator Marconi conducted his first experiments at the close of the last century, and there are those still living in the town who remember him working in their midst. There is not a wireless listener in Bournemouth or anywhere else who has not lost a good friend in Senator the Marchese Marconi whose death in Rome this week has been universally mourned.'

Madeira House Plaque

Madeira House, 2012
Looking toward Bournemouth Pier

Madeira House, 2012
Rear View

Madeira House, 2012
Reputed aerial mast guy wire mount from
original station

Madeira House, 2012
Main Entrance

The Haven Hotel

The first hotel built on this location was the North Haven Inn in 1838. The Inn was demolished and replaced with the present Haven Hotel in 1887. The Haven Hotel housed Belgian refugees during the First World War and was a military contact point during the Second World War and at one stage housed a naval detachment.

The Marconi Company continued to use the Haven Hotel experimental station until 1926. Marconi spent eight years living at the Haven Hotel and experimenting with wireless telegraphy between the Alum Bay and Niton stations on the Isle of Wight. The aerial masts were erected on the 'water' side of the hotel although in 1904 another smaller mast was erected at the back of the hotel. The first mast was then added to the smaller to make a complete aerial system just over 158 feet in height. This mast was dismantled in 1913 to allow builders to extend the hotel.

Marconi on the Isle of Wight
Cowes, 1923

Marconi often returned to the area, usually for the Cowes regatta, anchoring his 900 *Elettra* to the South of Brownsea Island by the castle. In August 1927, as a member of the Royal Yacht Squadron, Marconi took his daughter, Maria Cristina to the regatta at Cowes. Aboard the *Elettra*, the only yacht flying both the white ensign of the Royal Navy and the Italian tricolour with the crown of Savoy, they entertained royalty and nobility, tycoons and Hollywood film stars.

Today the Haven Hotel building has changed markedly from the 1901 photographs as it was partially rebuilt in 1926. In 1976, the Haven was purchased by the FJB Hotel chain and again altered with the introduction of a large sports and leisure centre including squash courts, saunas, steam and games rooms. The hotel can be found by following the signposts to Sandbanks through Bournemouth and then following the coast road into the one way system. As this road appears to lead away from the sea, the Haven Hotel lies on the left. The hotel maintains its connection with Marconi's work with a plush Marconi lounge containing two roaring fires and a group of captioned photographs on the wall. Over one of the fireplaces a small plaque records its piece of history.

'In this room which may truly be called the cradle of wireless, Guglielmo Marconi during the years from 1898 until 1926 conducted some of his most important experiments in wireless telegraphy and telephony and laid the secure foundations of a science of inestimable value to humanity.'

Haven Hotel Station from the sea – Poole, 1997

Haven Hotel Station – Poole, c. 1930

Haven Hotel Station – Poole – Interior Marconi Room

Haven Hotel Station – Poole – Plaque and Entrance to Marconi Room

Culver Cliff

Another 'Marconi' wireless station is recorded on the Isle of Wight. This was located high on Culver Cliff above the town of Bembridge on the Island's south eastern coast, sitting in the shadow of the impressive Yarborough monument. The 75 foot high obelisk is dedicated to Charles Anderson Pelham, Earl of Yarborough and Baron Worsley of Appuldurcombe in the Isle of Wight who died on 5th September 1846 aged 65.

The wireless station that stood on Culver Cliff has for many years been known locally as the 'Marconi Station'. The Culver Cliff wireless station was built sometime in late 1900, making it one of the earliest permanent wireless stations in the world and one of the first to be equipped with Guglielmo Marconi's new wireless telegraphy communication system. It was built by the Royal Navy and supplied with equipment from the very first contract for wireless equipment for 32 units after the successful trials during Royal Navy manoeuvres in July 1899. The equipment was supplied by the Marconi Company and manufactured at the worlds' first wireless factory, the Hall Street Works in Chelmsford, Essex. Delivery was completed by 31st December. Each wireless set had to pass an acceptance test by successfully transmitting between a ship, HMS *Minotaur*, in Portland harbour and another in Portsmouth harbour, and also with an intermediate ship up to a distance of 30 miles. All the sets successfully passed.

The Royal Navy's other stations were Malta, Gibraltar, Dover, Rame Head Scilly, Roche's Point and Portland. Although primarily naval wireless stations, Culver Cliff and its sister stations also doubled as coastguard stations. Culver Cliff would have been manned by Naval Signalman who would have received a course of instruction on one of the very first wireless training courses run at HMS *Vernon,* the torpedo training school ship based at Portsmouth.

Eight more coastal stations were planned for the financial year 1903/04. These were Bere Island, Spurn Head, Alderney, St. Abb's Head, St. Ann's Head, Landguard, Port Patrick and Duncansby Head. By 1906, using call sign 'CC' the Culver Cliff station was listed as performing 'Government work', or Royal Navy communications, operating on a wavelength of 350-400m and by 1910 its call sign had changed to 'RQN'. The station was still listed as being in use in 1920 by the British Admiralty.

A popular story concerning the Culver Cliff wireless station was that it was the first shore station to receive wireless distress signals from the doomed liner *Titanic* when it struck an iceberg on Sunday 14th April 1912. There is no evidence that this actually occurred and for the time it would have been a remarkable if not impossible technical feat. It is widely known that the *Titanic's* distress calls were picked up by wireless operators on ships up to 153 miles away which led to many lives being saved. The *Titanic* was also heard by a Newfoundland shore station but it is unlikely that with the equipment on board the *Titanic* that any signals would have carried across the Atlantic to the Isle of Wight.

All sign of the wireless station was lost by the building of the Culver Tavern some 35 years ago although the round shaped coastguard signal hut survived as the entrance porch until the infamous gales of October 1987.

Culver Cliff provides a spectacular walk and vantage point over the south western corner of the Island. The headland also has a privately owned fort built by the Palmerston administration as protection against Napoleon III between 1862-1867 for the huge cost of £48,025. The fort served in both world wars and the grounds were handed over to the National Trust in 1967. Culver Cliff also has a headland battery built in 1902. The gun emplacements were built to protect the Spit-head approaches with two 9.2 inch breech loading guns through two world wars but they are now half filled in for safety.

Culver Cliff – circa 1915
Showing coastguard wireless station

Culver Cliff – 1912

In front of the wooden building on the extreme right, with its own water tank and 150 feet high sectioned pine wood aerial mast was the coastguard station

Culver Cliff – 1997

Culver Cliff – circa 1988

Portland Bill

The Isle of Portland is a limestone island, 4 miles long by 1.5 miles wide, in the English Channel. Portland is five miles south of the resort of Weymouth and forms the southernmost point of the county of Dorset.

The Island of Portland had long been a naval base and Portland Harbour was formed by the construction of gigantic breakwaters. Before that the natural anchorage provided by the Bill had hosted ships of the Royal Navy for more than 500 years.

This long tradition of military use makes it inevitable that Marconi wireless equipment played a major part in operations on Portland, indeed an experimental Marconi wireless station was built on the Island, listed in the Annual Report of the Torpedo School for 1901. It is shown as having a Marconi set and was part of the Portland Shore Station of the Royal Navy. The station was actually established at the end of 1900 and like Culver Cliff was one of the first to receive Marconi wireless equipment from the first order of 32 units placed on the Company by the

Royal Navy. The Portland station was a little delayed as its equipment was used aboard HMS *Minotaur* as part of the acceptance trials for each of the new sets of equipment, tested to a ship moored in Portsmouth harbour and another at sea, up to a distance of 30 miles.

The Southern Times and The Portland Year Book for 1904/5 records that the 'Marconi Wireless Telegraph Station' was removed from near the Coastguard Station to the High Lighthouse in April 1904. *The Wireless-Telegraph Shore Stations of the World* list of 1st October 1906 shows Portland Bill operational as a British Admiralty station with Marconi equipment operating on 340m for 'Government use.'

The Southern Times also reported on 16th March 1907 that the Station had been in continual communication with Gibraltar for a week. In 1910 the range of the British Admiralty station located at Windmill Hill on Gibraltar, call sign SMP was only quoted as 90 nautical miles. In 1910 the British Admiralty station on Portland is listed as having the call sign TWQ.

In 1982 when I joined the Marconi Company as junior software engineer I was assigned to work on a Royal Navy communication project. One of the many transmitter sites was located on Portland Bill, where I led the project team that installed the Marconi ICS3 1kW powerbanks, the MFT H1141 10kW amplifiers, transmitters, aerials and all the associated command and control systems and software.

On Portland we also struggled to get a successful earth connection and at one stage had to resort to hanging coax cable over the cliff into the sea; just as George Kemp had done at other stations, and perhaps even at Portland, nearly 90 years before. In this way the Marconi Company kept its association with the island until the Royal Navy base closed down in 1995.

Portland Bill Lighthouse and Wireless Station

Portland Bill Lighthouse
and Wireless Station, c. 1911

Marconi ICS3 Amplifier Power Bank,
UKMACCS project. Portland Bill, c.
1986

Eaglehurst and Luttrell's Tower

Eaglehurst is a large rectangular house located not far from Calshot and Southampton, in the parish of Fawley, Hampshire. It is mostly a single storey building with a slate roof and has commanding views over The Solent to the Isle of Wight.

In the grounds of Eaglehurst there is a folly known as Luttrell's Tower, which was named after Simon Temple Luttrell who was of Irish blood and a former owner of Eaglehurst. The Tower, sometimes also known as Eaglehurst Tower, was built circa 1780 and is a large yellow bricked rectangular building of three storeys plus a cellar, together with a lead roof and a parapet.

In 1778 Luttrell married the daughter of Sir Henry Gould, and in 1793 he was arrested in Bologna by revolutionaries who mistook him for the brother of the King of England. He was released two years later and died in 1803. He had bought the land in 1772 for duck hunting and the tradition is that he built the tower for smuggling. It more likely started out as a grand hunting lodge, although the smuggling legend probably grew up from the network of tunnels that were dug underneath the tower, one of which still leads down to the beach. The passage of time and alterations have meant that the tunnel now no longer opens out directly on to the beach but on to a small patio slightly raised from the beach.

Queen Victoria, while still a Princess, visited the estate in 1833 aged 14. During the visit she climbed the tower and wrote in her diary that there were fine views of Norris Castle on the Isle of Wight. She also commented that, at that time, one of the rooms in the tower was home to an Egyptian Mummy, from which she was given a piece of the linen in which it was wrapped. It is said that Queen Victoria considered buying the property when it came up for sale but eventually chose Osborne House on the Island instead. A stone plaque, fixed to the base of the tower, bears the following inaccurate legend:

> 'Luttrell's Tower built 1730. Here Marconi concluded his wireless experiments during the great war of 1914 – 1918.'

Looking for a 'place in the country', Guglielmo Marconi's Irish wife, Beatrice, found Eaglehurst, which included Luttrell's Tower, and the premises were let to the Italian inventor's family in 1911. Marconi used the top room of the tower as a laboratory. Marconi's daughter Degna wrote that as youngsters what they enjoyed about Eaglehurst, besides their pony cart and the sheltered, pebbly beach, was the tower. 'This was a curious and entertaining eighteenth century architectural folly, a narrow three-story structure - crowned with a round turret surmounted by the inevitable battlements and a flag flying.'

In 1912, Marconi and his family were invited by the White Star Line's Bruce Ismay to be his guest on the maiden voyage of the *Titanic*. Fortunately for Marconi he had to hurry to America on business, so he went earlier on the RMS *Lusitania*. Travelling with the then Managing Director of the Marconi Company, Godfrey C. Isaacs and his son Marcel, Marconi had arrived in New York on 15th March. He had to be in America as the Marconi Company had brought a legal action against the United Wireless Company for patent infringement that was being heard in the Federal Court on 25th March.

Marconi's wife Beatrice retained her booking, but on the eve of the voyage, she too cancelled due to their son Giulio's sudden illness with a high fever. She cabled Marconi that she had to postpone her trip and settled down to watch over her youngest and to face another of the endless separations that so disrupted her marriage. The *Titanic* sailed on her fateful voyage on 10th April. Marconi still held a ticket for the *Titanic's* return voyage to England on 20th April 1912.

Degna, his eldest daughter, recalled how she and her mother Beatrice climbed the tower on the lawn above the water's edge on the morning of 10th April to watch the *Titanic* sail off on its maiden voyage. Degna was not yet four years old, and yet she still recalled how tightly her mother held her hand and she sensed that she was sad. When she was older she knew why, her mother had wished she were on board sailing to America to meet her father. Degna remembered vividly:

> .'...together we waved at the ship, huge and resplendent in the spring sunlight, and dozens of handkerchiefs and scarves were waved back at us. As the Titanic passed from our view over the calm water, we slowly descended the steps. It was a long way down....'

Degna wrote: 'The departure saddened Beatrice. She had wanted very much to be aboard.

Guglielmo arrived in New York just in time to hear that a wireless message had been received at the Cape Race wireless station in Newfoundland which might indicate a disaster at sea. The *New York Times* promptly sent a wireless message to the *Titanic's* Captain, Edward J. Smith, Master for all White Star maiden voyages, who was due to retire after this one, but they got no reply.

A period of total confusion ensued. The full horror and tragedy of the disaster was only fully comprehended when the *Carpathia* came up New York Harbour through the rain on Thursday night. As soon as her gangplank went down, Guglielmo Marconi stepped out of the immense and silent crowd at Pier 54 and, with police clearing the way, was one of the first, with Mr. Speers of *The New York Times,* to go aboard to interview the Marconi Company wireless operators whose work had saved so many.

Marconi met with the *Carpathia's* operator Thomas Cottam and *Titanic's* second wireless officer, Harold Bride who had been severely injured. The ship's first wireless officer, John George Phillips, had drowned. Marconi later gave evidence at the *Titanic* enquiry. The principal finding was inescapable. Without wireless on board the *Titanic*, all 2,224 passengers and crew would have perished.

The tragedy of the sinking of the 'unsinkable' liner had thus nearly destroyed the Marconi family, but Marconi's radio system revolutionised safety at sea, in part driven on by the *Titanic* disaster.

For Marconi personally the First World War was to be a turning point in his life and career. In July 1914, at Buckingham Palace, King George V had awarded Guglielmo Marconi an honorary knighthood. This was an act of grateful recognition of his huge contribution to the new science of wireless but also, perhaps, a tacit apology for the harm done to his name by innuendoes of financial scandal two years before.

Later that month, at Spithead, Marconi and his wife Beatrice lunched on one of the Royal Navy battleships anchored for the annual review. Next morning the ships were gone, and within days, as a citizen of Italy, still neutral in spite of its 'triple alliance' with Austria and Germany, Marconi became an 'alien' and the subject of suspicion. For a while, down by the Solent at Eaglehurst, his country house, he was treated by locals as an enemy spy. He remained there with his family until, in the words of his daughter Degna, 'sanity reasserted itself'.

Marconi resumed travelling, between Britain, the USA and Italy, where, as a non-combatant, he was elevated to the Senate in Rome in January 1915. In order to give evidence in the on-going commercial and legal battle with Telefunken in the US courts, in April 1915 Marconi had to sail from Southampton to New York on board the SS *Lusitania*.

On her return journey without him, on 7th May 1915 the ship was torpedoed and sank in just 18 minutes, killing 1,198 of the 1,959 people aboard, leaving just 761 survivors. Some papers even speculated that it was an attempt to kill or capture Marconi. The sinking turned public opinion in many countries against Germany and eventually contributed to America's entry into World War I. The loss of the *Lusitania* became an iconic symbol in military recruiting campaigns as to why the war was being fought.

Marconi's youngest daughter, Gioia, was christened at Fawley church in 1915, and the family left Eaglehurst for Italy in the same year. The war had changed everything and they never returned.

Luttrells' Tower

Marconi with his
Children at Luttrells

APPENDIX TWO

Island Communication,
Cables
and
Conduction

On 14th October 1853, the Isle of Wight was connected to the mainland by an early telegraph cable and its associated land lines, opened from Southampton to Osborne House, the Queen's residence on the Isle of Wight. It worked in concert with the circuits of the Electric Telegraph Company. The construction was unusual in that it comprised a cable from Keyhaven to Hurst Castle buried three feet under sea mud along a long spit, and a true underwater cable from Hurst Castle to Sconce Point, one-and-three-quarter miles in length. There was another underwater cable on the line under the River Yar at Yarmouth. The cable utilised Charles West's india-rubber insulation and "plaited iron wire" armour, and was laid by the International Telegraph Company's cable vessel, the 530 ton *Monarch*. The busy nature of the sea-way leading out of Southampton to the Atlantic was such that the cable needed continual repair from anchor damage. The Electric Telegraph Company laid a second cable from Hurst Castle to Sconce Point during 1867.

In July 1877 William Preece brought over from America the first pair of practical telephones seen in Great Britain. Later in the same year Bell's perfected type of telephone was exhibited at a meeting of the British Association in Plymouth. However the telephone hit the papers in January 1878 when Alexander Graham Bell demonstrated his new telephone invention to a curious Queen Victoria on 14th January at Osborne House, East Cowes in the Council Room with calls placed to London, Cowes and Southampton. These were the first long distance calls in the UK.

Bell, with the support of the Post Office, had connected telephones between the house and Osborne Cottage in the grounds of the Queens residence. After a brief lecture on the science, Bell called the cottage and the Queen had the first royal conversation on the telephone. Calls were then made to Cowes, and using this telegraph cable, to Southampton and London on the mainland. Unfortunately the line from Cowes failed at the crucial moment but calls were successfully made between the Cottage and Osborne House. The Queen stated: 'It is rather faint and one must hold the tube rather close to one's ear.'

The Telephone Company Ltd (Bell's Patents) was formed to market Bell's patent telephones in Great Britain. The Company was registered on 14th June with a capital of £100,000. Its premises were at 36 Coleman Street. It had a capacity for 150 lines and opened with 7 or 8 subscribers. One of the first telephone lines to be erected in the vicinity of London was from Hay's Wharf, south of the Thames, to Hay's Wharf Office on the north bank.

The first trial of long distance telephony in Great Britain as a commercial proposition was held on 1st November with a call between Cannon Street in London and Norwich, a distance of 115 miles, using an Edison transmitter on a telegraph wire. William Preece from the Post Office Engineering staff was asked whether the telephone would be an instrument of the future which would be largely taken up by the public. He replied: 'I think not'.

Questioned further he said:

> 'I fancy the descriptions we get of its use in America are a little exaggerated; but there are conditions in America which necessitate the use of instruments of this kind more than here. Here we have a superabundance of messengers, errand boys, and things of that kind.'

Another pretender to the throne of long distance communication was the electromagnetic induction telegraph system sponsored and developed by the same William Preece.

For more than 40 years, as a lifetime employee of the Post Office, Preece worked on his system of wireless telegraphy by induction. Inductive telegraphy, a system by which signals could be exchanged between two locations, separated, perhaps, by a body of water, involved the laying out of long wires in each location parallel to each other. Fluctuating currents flowing in one wire would then induce fluctuating currents in the other, although at greatly attenuated levels. There had been many experimenters with this method and Preece was probably one of the last of them, but he achieved more solid success than most.

The techniques involved were simple, particularly for a man with access to the manpower and equipment of the Post Office. Stringing the long wires that inductive telegraphy required was not, after all, very different from the jobs that telephone and telegraph linesmen did every day. It worked. Preece had the results to prove it.

In March 1882, Preece, now the Chief Engineer of the British Post Office, installed an induction based wireless communication system across the Solent, between the UK mainland and the Isle of Wight, when the submarine telegraph cable failed at Hurst Castle.

Preece wrote:

> 'For some cause the cable broke down, and it became of great importance to know if by any means we could communicate across, so I thought it a timely opportunity to test the ideas that had been promulgated by Prof. Trowbridge.'

Preece's system aimed to replace the land lines that existed between Portsmouth (Southsea), Southampton and Hurst Castle, and also those between Sconce Point, Newport and Ryde.

Morse signals were transmitted and received between Southampton and Newport with considerable success, using a telephone receiver, there not being enough current to operate a paper tape inker. When the cable was repaired and this method discontinued, some commented that the iron sheath of the broken cable had probably helped the results. Preece wrote:

> 'The Isle of Wight is a busy and important place, and the cable across

at Hurst Castle is of consequence. I put a plate of copper, about 6 feet square, in the sea at the end of the pier at Ryde. A wire (overhead) passed from there to Newport, and thence to the sea at Sconce Point, where I placed another copper plate. Opposite, at Hurst Castle, was a similar plate, connected with a wire which ran through Southampton to Portsmouth, and terminated in another plate in the sea at Southsea Pier. We have here a complete circuit, if we include the water, starting from Southampton to Southsea Pier, 28 miles; across the sea, 6 miles; Ryde through Newport to Sconce Point, 20 miles; across the water again, 1¼ mile; and Hurst Castle back to Southampton, 24 miles.'

'We first connected Gower-Bell loud-speaking telephones in the circuit, but we found conversation was impossible. Then we tried, at Southampton and Newport, what are called buzzers (Theiler's Sounders)--little instruments that make and break the current very rapidly with a buzzing sound, and for every vibration send a current into the circuit. With a buzzer, a Morse key, and 30 Leclanché cells at Southampton, it was quite possible to hear the Morse signals in a telephone at Newport, and vice versa. Next day the cable was repaired, so that further experiment was unnecessary.'

The instruments used were a telephone in one circuit, and in the other about twenty-five Leclanché cells and an interrupter. The sound could then be heard distinctly and so communication was kept up until the cable was again in working order. Of the two lines used in this case, one extended from the sea at the end of the island near Hurst Castle, through the length of the island, and entered the sea again at Ryde; while the line on the mainland ran from Hurst Castle, where it was connected with the sea, through Southampton to Portsmouth, where it again entered the sea. The distance between the two terminals at Hurst Castle was about one mile, while that between the terminals at Portsmouth and Ryde amounted to six miles.

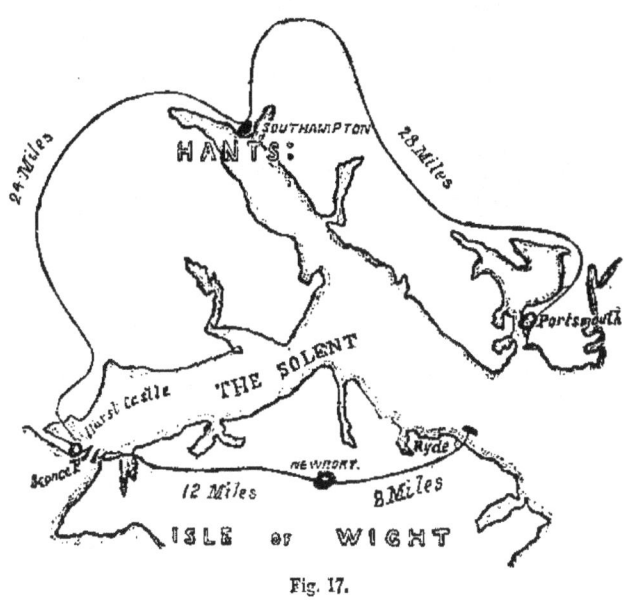

Fig. 17.

Preece's Isle of Wight induction system

William Preece

In 1892, William Preece signalled between two points on the Bristol Channel, and across Loch Ness in Scotland, employing both induction and conduction to affect one circuit by the current flowing in another.

In March 1898 Preece's induction system was permanently established for signalling between Lavernock Point and the Flat Holm and was handed over to the War Office. Permanent lines of heavy copper wire were erected parallel to each other, one being on the Flat Holm and the other on the mainland.

In 1899, he had also transmitted across the Menai Straits, a distance of nearly half a mile, actually sending messages more rapidly than the new Marconi wireless system but less distinctly. In 1900, Preece was sending signals nine miles from Lavernock to Weston-super-Mare, and 25 miles from Ilfracombe to the Mumbles.

In reality the induction system did not work very well. There were some situations in which it did not work at all. The greater the distance over which signals had to be transmitted, the longer the parallel wires had to be at each location. Preece's rule of thumb was that each wire had to be as long as the distance between the stations so to signal over 100 miles by inductive telegraphy, you would require parallel wires at each location 100 miles long. If this was awkward enough in normal situations the problems of communicating with a small island, or a lighthouse, perched on some remote rock were insurmountable.

It was clear to Preece that his system was never going to be practical for communication with ships and that he had already reached a practical limit on range and reliability. Inductive telegraphy worked best when there was lots of land area over which to lay out the wires. This was, however, just the situation in which conventional wired telegraphy worked best, and with no greater cost in wire.

The history of technology is full of dead ends and blind alleys. They attract less attention than the open roads, but it is important that we remember those who tried and what they achieved. History also prefers success to failure, achievement to frustration.

In 1896 William Preece found himself at a technological dead end and it says much for his intelligence that he recognised the fact. It also speaks well for his ability to swallow his pride, for the decision to cut his losses and take up with Marconi cannot have been an easy one.

William Preece, during his brief honeymoon with Marconi, would insist on basking in the reflected glory of having 'discovered' the Italian inventor, and continued to lecture to audiences around the country on the great value this new sort of wireless telegraphy might have for lightships and lighthouses.

But with Marconi's work the world was about to change forever.

APPENDIX THREE

Description of Alum Bay Operations
7ᵗʰ May 1898

George Kemp's diary for 1898 records that:

> April 26ᵗʰ – Went over to take charge of the Needles Station.

> May 7ᵗʰ – Gave a demonstration to Admiral Sir M.C. Seymour who was relieving Captain A.W.A. Wood in command of 'The Monarch' in Besika Bay, Admiral Lord C. Beresford, Lieut. Hornby and others, four in number, from the Royal Navy.

The following is part of the report that Hornby wrote for the Royal Navy for their annual review paper for 1898. It provides an excellent description of Marconi's Alum Bay station located in the Royal Needles Hotel, May 1898.

EXTRACTS FROM A REPORT, BY COMMANDER HORNBY, ON SOME EXPERIMENTS WITH WIRELESS TELEGRAPHY, CARRIED OUT AT THE NEEDLES, BY M. MARCONI ON THE 7ᵀᴴ MAY 1898.

The experiments carried out on this occasion were merely calculated to show the feasibility of transmitting invisible signals from one station to another at a considerable distance, without the aid of a connecting wire, where by means of a suitable apparatus they could be received and recorded on an ordinary Morse telegraph recorder. In this M. Marconi was entirely successful, the signals sent from Bournemouth being received with remarkable clearness at Alum Bay, a distance of 14 ½ miles, and v.v.; the speed of signalling being about 10 words a minute.

The method of working may be briefly explained as follows: The sending apparatus consists of an ordinary Ruhmkorff induction coil, capable of giving a 10 inch spark, joined up to a battery with a key in the circuit, by means of which the circuit can be made and broken in practically the same manner as with an ordinary telegraph key. The two ends of the secondary wire of the coil are joined to two ¾ inch brass balls about 1 inch apart, to one of which is also attached a long insulated wire, or wires, tied up to a flagstaff.

On the present occasion the wire consisted of three parts arranged in a ladder form with wooden rungs. The other ball is joined to earth. Whenever the key is pressed the coil is brought into action, and sparks pass continually between the balls. These sparks are the expression of oscillatory currents produced by the coil, the rate of oscillation being estimated under ordinary circumstances at about 30 million per second. These currents also surge up and down the vertical wire, and each one propels a wave or undulation radially outward through the ether, somewhat in the same way that ripples spring outward when a stone is dropped into still water.

These undulations travel away through space in much the same way as light or sound waves, being, moreover, reflected by metal surfaces on which they impinge. Also if they encounter any vertical conductor of electricity, they generate in it a series of definite electrical pressures tending to force currents of electricity along the conductor, and if this is in connection with a suitable receiving arrangement the receiver will be actuated by these currents and the signals recorded.

The receiving apparatus consists of a glass or ivory tube known as a 'coherer', in which are two silver plugs or balls, about 1/4 inch apart, the space between them being filled with nickel and silver filings. Wires are led in at either end of the tube, and to one is attached the vertical receiving wire, and to the other a wire connected to earth.

The two wires from the coherer are also joined through two choking coils of wire to a single cell and a telegraph relay.

Under ordinary conditions the electrical resistance of the coherer is so great that the single cell cannot force sufficient current through the relay to work it; but directly one of the currents induced in the receiving wire reaches the coherer it causes its electrical resistance to drop greatly, sufficiently so to allow the single coil to work the relay. If the coherer is tapped or shaken mechanically it reverts to its original condition, and is ready for the next wave.

The distance to which signals can be transmitted and received depends mainly on the vertical height of the wires, 30 feet being allowed for 1 mile and the distance varying as the square of the height (i.e., 60 feet gives 4 miles, and 120 feet gives 16 miles). This may be accounted for by considering that the undulations are imparted to the surrounding medium from each minute length of the vertical wire, thus the greater its length the greater will be the disturbance propelled through the atmosphere. Similarly, the greater the vertical length of the receiving wire the larger will be the interception by it of the undulations, and the greater, therefore, the electrical effect produced in it by them.

At each station a set of sending and receiving apparatus are necessary, the receiving apparatus (except the recorder) being enclosed in an iron box, to screen it from the effects of the sending coil, etc., which would otherwise damage it.

The same vertical wire can be used to send or receive, being shifted over from transmitter to receiver as required.

At present the chief difficulties in the employment of this method of signalling seem to be:

1. Only one instrument can be sending at a time, and this will actuate all receivers within its range. A second instrument trying to send at the same time would also affect all receivers and produce an unintelligible jumble of the signs. Signor Marconi professes to have discovered a method by which he can attune two transmitters and receivers, and that instruments so arranged will only receive signals from the transmitter which is in tune with them, and will be totally unaffected by other instruments sending in its vicinity. This he effects by means of a coil of wire containing an exactly equal number of turns inserted in both sending and receiving wires. This arrangement was unfortunately not shown on the occasion of the visit to Alum Bay.

2. Difficulty of obtaining sufficient height for the masthead wire. Signor Marconi proposes to carry out experiments with balloons or kites to overcome this difficulty, and of course in ships a considerable vertical height is generally obtainable. In the *Defiance* it has been found advisable for ship work to run a wire down from the masthead to each equipment, thus obviating the screening effect of funnels, etc.

3. Preserving the insulation of the sending coil in wet weather, sparks being apt to fly about, and nasty shocks received by the operators. This is not insurmountable, but would necessitate the apparatus being adequately housed in a cabin or other well protected place, and Signor Marconi has found that signals are actually better transmitted on a foggy day than on a fine one.

4. Sufficiently delicate adjustment of the receiving apparatus, especially the relay, and also the de-cohering of the tube by the tapper.

Signor Marconi stated that the apparatus had been working untouched for five weeks, and that once adjusted the apparatus should remain correct for a considerable time.

Undoubtedly this method of signalling has great possibilities, and its development by Marconi should be carefully watched, especially if his claim is substantiated of being able to confine his signals to certain instruments, the signals being totally unaffected by other instruments not absolutely in accord with them.

The advantages which this method of signalling would possess for naval purposes are clearly summarised in Captain H.B. Jackson's report given at page 100 of A.R. (Admiralty Report) 1897. From recent reports from *Defiance,* it appears probable that a mechanical tapper will be introduced, thus simplifying the electrical adjustments, and increasing the speed and certainty of signalling.

The coherer is partially filled with metallic dust (96 per cent. nickel and 4 per cent. silver), and the air exhausted, the vacuum, however, being not important.

The form of the filings, however, is important, those of a jagged or sharp form being the most sensitive. They are selected under the microscope, and should be as nearly as possible of a small size. The amount per tube is only about 20 to 25 grains.

APPENDIX FOUR

Description of a wireless operator's room

Today we cannot conceive of what an early wireless room was like. The noise, smell, smoke, gas and fumes that were generated by early spark transmitters made each room a cacophony of flashing lights and sparks that cracked and hissed. This was what Marconi converted the Alum Bay Needles Hotel billiard room, Niton farm house, and the Madeira House and Haven Hotels rooms into. It was perhaps amazing he managed to stay so long at each place.

In 1904 a wireless operator was described:

'In his shirt-sleeves at the big key, rapping out a call that, within the narrow confines of a little cabin, sounded like sputtering pistol-shots; showing blue-white lightning flashes as the current leaped from the 'sparker' at each bend of the wrist, and causing blue flames to play about the six Leyden jars and at the contacts of his key.'

The room had a characteristic smell, an odour of gas, ozone, acid, hot rubber insulation, heat and smoke. Any electric lights in the room would dance to the rhythm of the Morse code key being pressed. The risk of electric shock of any metal fittings in the room or nearby bathrooms was a constant risk, touching the coil or spark balls would be instantly fatal and getting near the aerial wire would give a severe burn and shock.

'For minutes at a time the call shot forth, then the operator would shift connections to the receiver and listen for as many minutes, carefully adjusting and readjusting the delicate instruments meanwhile. There were two of these receivers, one to take the place of the other if a fault should show while a message was coming in. They were blocks of delicate and complicated machinery, carefully cased in wooden boxes, their supports carefully padded to steady them against the least vibration. There was a strip of paper to record messages, like the tape on the old Morse telegraph instruments; but that is not necessary to the operator who can 'read by sound.' This accomplishment is no mean one, however, for the sound is a very delicate ticking hardly to be noticed by the inexpert, and very different from the pistol-shots of the sender.'

There was one other notable characteristic about the spark gap transmitters at this time. On reception, each signal sounded just a little bit different from the rest as every spark station had its own characteristic sound. The received signal hissed it Morse code dots and dashes while the mechanical Morse code inker clicked and whirled away recording the signals.

Guglielmo Marconi, 1901

GLOSSARY

Aether: An older spelling of ether, now not commonly used. See ether.

Aerial: A device which emits and receives radio waves. If you were a German or French you would use *antenne*, if a Spaniard, *antena* and if an Italian, *antenna*. America followed the Italians and adopted the use of the word *Antenna*, while Britain, especially inside the Royal Navy always used the term aerial. Also see antenna.

Aerial lead-in: A lead-in from the aerial (or 'antenna') to the main device.

Alternator: An Alternating Current (A.C.) electrical generator.

Antenna: Identical usage to the word *aerial*. The part of a radio system that is used to radiate radio waves into free space or to extract energy from incoming radio waves. An assembly of simple antennas arranged in space so that it has some degree of directionality is known as an antenna array.

Battery: A portable device to supply electric potential and current. An electrical battery is a combination of one or more electrochemical cells, used to convert stored chemical energy into electrical energy. Since the invention of the first voltaic pile in 1800 by Alessandro Volta, the battery has become a common power source for many household and industrial applications. Batteries may be used once and discarded, sometimes known as a 'dry cell' or recharged for years.

The name 'battery' was coined by Benjamin Franklin for an arrangement of multiple Leyden jars (an early type of capacitor) after a battery of cannon. Strictly, a battery is a collection of two or more cells, but in popular usage battery often refers to a single electrical cell.

Branly Coherer: One of the original types of coherer radio detectors developed by French scientist Edouard Branly in 1890. See also Coherer.

Capacitance: A unit measurement of a system's ability to store electrical charge.

Capacitor: A device used to store and release electricity, sometimes as the result of a chemical action. Also referred to as a storage cell, a secondary cell or a condenser. A Leyden Jar is an early example of a capacitor.

Coherer: An early form of signal detector in wireless telegraphy, developed by Branly, Lodge, Marconi and others. It is based around the effect that small particles of metal filings stick together (or 'cohere') when an electric field is present. A coherer circuit consisted of a basic electromagnetic wave detector for various wavelengths and a circuit that obtained signals from modulated radio waves. The coherer then 'decoded' these signals.

The operation of the coherer is based upon the large resistance offered to the passage of electric current by loose metal filings, which decreases under the influence of radio frequency alternating current. The coherer became the basis for radio reception, and remained in widespread use for about ten years. It was used by Guglielmo Marconi, in his early experiments. Oliver Joseph Lodge improved Edouard Branly's coherer as a detector of wireless waves by adding a *'trembler'* or 'tapper' which mechanically dislodged the clumped filings, thus restoring the device's sensitivity and allowing the development of a continuous radio detector

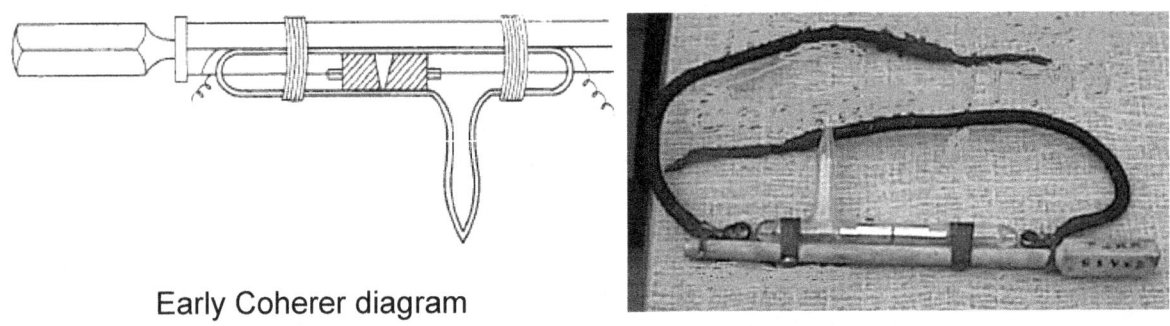

Early Coherer diagram

Early Marconi coherer

The instant an electric discharge from a spark transmitter occurred in the vicinity of the receiver the coherer's metallic filings 'cohered' or stuck together and become highly conductive. Unless the cell was physically disturbed or jarred these granules would continue to adhere to each other and the device would continue to conduct. But once the device was tapped lightly its conductivity vanished and it became an insulator.

The addition of a mechanical trembler, developed by Oliver Lodge (who also added other technical improvements), activated by local battery, was a simple solution which required no intricate timing system, since it made little difference if the cell was continuously tapped. When connected in a low voltage circuit, in its ordinary state a coherer had a resistance of millions of ohms, but this dropped dramatically to hundreds of ohms when electromagnetic waves were produced in the vicinity. Basically, the coherer acted like a voltage-controlled switch that closed when a wireless signal was received.

Marconi and his engineering team spent thousands of hours developing and refining the coherer design and experimenting with every conceivable mixture of metallic dusts and powders to fill the tube.

The invention of the coherer spurred the development of wireless telegraphy and it was widely used from the early 1890s until about 1910.

Crystal: A crystalline piece of a semiconductor used in a wireless set as a detector of electromagnetic radiation on account of its properties of electrical conduction. Materials used include germanium, silicon and galena. Crystal detectors replaced coherers and magnetic detectors around 1910.

C.W: See Continuous wave.

D.C.: Direct Current: an electric current which flows in one direction only. The other system of electric current transmission is A.C.

Earth: The conducting mass of the earth or any conductor in direct connection with it, or at zero potential with respect to earth. It can also mean the connection (deliberate or accidental) between a conductor and the earth. Also known as 'ground.'

Electromagnetic Wave: Waves of associated electric and magnetic fields each at right angles to each other and to the direction of propagation. e.g., radio waves and light.

Ether: (Also traditionally spelt 'Aether') The science of radio was long established before the means by which radio waves actually travelled through the atmosphere was fully understood. The answer did not come from radio engineers, but from physicists' who were attempting to unravel the secrets of the atom. Before the science of quantum mechanics was determined it was thought that radio waves, like sound waves in air or even waves on a pond must travel in some medium, but nobody could ever find it. This all pervading, invisible fluid, in which the whole material universe was supposed to float was known as the ether, and was used for many years to explain the transmission of wireless waves. We now know that it was never really there to find, but it was a nice idea.

Frequency: The measurement of the number of times that an event occurs per unit of time. The standard unit of frequency is the hertz (Hz.).

General Post Office: (G.P.O) was officially established in England in 1660 by King Charles II and it eventually grew to combine the functions of both the state postal system and the state telecommunications carrier.

Originally, the G.P.O was a monopoly covering the despatch of items from a specific sender to a specific receiver. The postal service was known as the Royal Mail because it was built on the distribution system for royal and government documents. In 1661 the office of Post Master General was created to oversee the G.P.O.

When new forms of communication came into existence in the 19th and early 20th centuries the G.P.O claimed monopoly rights on the basis that like the postal service they involved delivery from a sender and to a receiver. This was used to expand state control of the mail service into every form of electronic communication possible on the basis that every sender used some form of distribution service. These distribution services were considered in law as forms of electronic Post Offices. This also applied to telegraph and telephone switching stations.
In the mid 19th century several private telegraph companies were established in the

UK. The 1868 Telegraph Act granted the Postmaster General the right to acquire inland telegraph companies in the United Kingdom and the Telegraph Act of 1869 conferred on the Postmaster-General a monopoly in telegraphic communication in the UK. Overseas telegraphs did not fall within the monopoly. The private telegraph companies were bought out. The new combined telegraph service had 1,058 telegraph offices in towns and cities and 1,874 offices at railway stations. In 1869, 6,830,812 telegrams were transmitted, producing a revenue of £550,000.

The same principles were applied to telephone, wireless telegraph and wireless telephone services. This latter expansion then incorporated wireless broadcasting which was non-specific in terms of delivery from sender to receiver. At first the G.P.O referred to all broadcasting transmitters as senders, while individual receivers retained that name. Like the mail, everything was licensed by the General Post Office under the terms of its Royal Charter This meant that the G.P.O maintained a monopoly on all communications into, out of, and within the British Isles.

Ground: See Earth.

Ground Plane: A large or significant mass that presents the effect of earth (ground) to an aerial system.

Hertz, Heinrich Rudolf: (22nd February 1857 – 1st January 1894) was the famous German physicist who confirmed in the laboratory James Clerk Maxwell's work on the electromagnetic theory of light.

In 1880 Hertz became an assistant professor at the Physics Institute of Berlin, working for von Helmholtz, who suggested he work on the need for a proof of Maxwell's theory about the velocity of electromagnetic forces. Hertz studied Maxwell's theory, and appreciated the need for generating short wavelengths and the problems in trying to do so. After a period of about two years at Kiel University, Hertz moved to the Technical High School at Karlsruhe. The Karlsruhe laboratory was particularly well equipped. By his own account, at this laboratory he found and used for lecture purposes a pair of so-called Riess or Knochenhauer spirals essentially a flat, spiral coil.

He noticed that sparks across one could induce, without direct connection, sparks in the other coil. He studied the effect systematically, leading to a paper in 1887: 'On Very Rapid Oscillations.'

Heinrich Hertz was the first to demonstrate the existence of radio waves. In 1888, in a corner of his physics classroom at the Karlsruhe Polytechnic in Berlin, Hertz generated electric waves using an electric circuit. The circuit contained two metal rods separated by a short gap, and when sparks triggered by a high-voltage electrical capacitive spark discharge crossed this gap strong oscillations of high frequency were produced. Hertz proved that these waves were transmitted through air by detecting them with some distance away with a simple wire loop. A small spark in this receiver gap signified detection of the electromagnetic waves.

Hertz discovered the progressive propagation of electromagnetic action through space, and measured their length and speed. He also showed that like light waves they were reflected and refracted. Hertz also noted that electrical conductors reflect the waves and that they could be focused by concave reflectors. He found that non-conductors allowed most of the waves to pass through. What Hertz had discovered were the longest waves of the electromagnetic spectrum that includes gamma, X-rays, visible light, infrared, ultraviolet and mic

All of these findings were first published in 1888 the journal *Annalen der Physik,* then in Hertz's first book, *Untersuchungen Ueber Die Ausbreitung Der Elektrischen Kraft* (Investigations on the Propagation of Electrical Energy (5th March and 7th May, 1888). His book is considered to be one of the most important works of science. This is where he first describes his confirmation of the existence of electromagnetic waves and how he was able to trigger his electromagnetic waves with his oscillator.

However Hertz failed to recognise that there might be any practical application to his work and he died at the very young age of 36. His discoveries would later be more fully understood by others and formed the foundation of the new wireless age. His experiments also became the basis of practically every early system of wireless communication including Marconi's.

Hertz's classic experiments also explain reflection, refraction, polarization, interference and even the velocity of his electric or 'Hertzian waves.' If anyone can be accorded the title of 'Father of wireless' it should be Hertz.

hertz: The standard unit of frequency, named after Heinrich Hertz, represented by the symbol Hz. One hertz is an event that repeats (or cycles) once per second, two hertz is an event that repeats (or cycles) twice per second, and so on.

Hertzian waves: Late 19th/early 20th century term for electromagnetic or radio waves; sometimes used to refer to early experimental wireless telegraphy. The waves were named after the German physicist **Heinrich Hertz** discoverer of wireless waves including Ultra High Frequency (U.H.F) and identified them as part of the electromagnetic spectrum.

H.F.: See R.F.

Hz: See hertz.

Inductance: The property of an electric circuit or device by which an electromotive force is induced in the circuit itself as the result of opposition to change in magnetic flux, or current flow.

Induction Coil: An apparatus for producing electric currents by induction. It is also referred to as a spark coil or Ruhmkorff coil, after Heinrich Daniel Ruhmkorff (Rühmkorff), a German instrument maker.

The induction coil is a type of electrical transformer used to produce high-voltage pulses from a low-voltage direct current (DC) supply and is the direct ancestor of the modern-day transformer and it works on the same principle as the spark coil in a car.

After the invention of the electric battery by Allisandro Volta in 1800, followed by experiments in electromagnetic induction by Michael Faraday and Joseph Henry in the 1820s, the first high-voltage induction coil was developed by Charles Page, a New England inventor.

Page's induction coil employed a primary coil of some turns of heavy copper around a bundle of soft iron rods and a secondary coil of many turns of fine wire wrapped on the outside of this primary coil. To interrupt the flow of current and cause a spark to jump across the terminals of the secondary coil, Page ran a screwdriver across a rough file between the battery and the induction coil. This crude method was greatly refined by other inventors and, by the 1850s, Ruhmkorff, a German living in Paris, had produced the first standard form of induction coil and gave his name to the device that would find its way into most of the physics laboratories of the later 19th century.

Because alternating-current systems were not available until the end of the 19th century, the induction coil was originally developed for use with direct current. These induction coil designs could take 6 to 12 volts from a series of chemical batteries and step up the voltage to many thousands of volts.

Induction coils were rated according to the length of their sparks, and many a laboratory prided itself on the length of the spark their induction coil could jump. By the late 1880s, quite a number of companies including Max Kohl, Radiguet, Leybold and the Newton Coil Company whose equipment was often used by Marconi, manufactured fine-quality induction coils for the high-voltage experiments, medical equipment and spark transmitters that constituted the cutting edge of physics research of the latter part of the 19th century.

The coil actually consists of a small primary coil and a large secondary coil where the high tension is formed by a mechanical breaker system. In the wooden base there is room for a capacitor made of paper and metal foil. Ruhmkorff or spark coils were also used to light the Geissler tubes and later the Crookes tubes that were used in the discovery of electrons by English physicist William Crookes.

Marconi's early equipment relied on 10inch and later 18 inch coils (the length of the spark when not connected to an aerial load) to generate the energy needed for the sparks in his transmitter.

Ink-Writer: More commonly referred to as a Morse Tape Inker. A device that physically inked or printed the dots and dashes of received Morse code onto a paper tape.

Insulator: An object or material that, by the property of high resistance, insulates the surrounding material from (usually high voltage) electricity. In more general terms, a non-conductor of electricity, sound or heat.

Jigger: A term applied to Marconi's important development of wireless tuning. A jigger was essentially a circuit component that enabled messages to be received from a selected station and filtered out interference from any other stations. It also concentrated the signal's energy into a narrower band, improving the overall performance of both transmitter and receiver. A jigger was essentially a high frequency transformer, whose primary circuit was connected to a Leyden jar (now called a capacitor), forming a resonant circuit which reduced the bandwidth of the signal to be transmitted coming from an induction coil. The signal then goes to the central coil (secondary circuit) whose terminals are connected to the transmitting aerial and the earth circuit, to be sent into the space. A similar tuning circuit, mounted inside the receiver, ensures the reception only of the signal coming from that transmitting station, ignoring all other signals present in the aerial.

kW: kilo watt used to denote output power of a transmitter. 1kW = 1,000 watts.

L.F.: See R.F.

Morse Tape Inker: A device for recording the Morse code (i.e., dots and dashes) of the received signal upon paper recording tape. It required a certain amount of signal power to operate. Used in electrical (wired) telegraphy and early wireless telegraphy.

M.F.: See R.F.

Obach Cell: A type of dry cell battery where the electrolyte fluid is mixed with some substance to form a paste. The battery is then sealed so it can be used in any position and is safe from spillage of the strong acid associated with lead-acid type batteries. This was an essential safety feature for ship mounted early wireless equipment.

Poldhu: Radio Station site in Cornwall, built by Marconi, which first transmitted the Morse code 'S' signal across the Atlantic in December 1901.

Post Master General: In the United Kingdom, the Postmaster General is a now defunct ministerial position. The King's letters to his subjects are known to have been carried by relays of couriers as long ago as the 15th century. In 1510, Sir Brian Tuke was appointed as 'Master of the King's Post.' In 1609 it was decreed that letters could only be carried and delivered by persons authorised by the Postmaster General. 1660 saw the establishment of the General Letter Office, this would later become the General Post Office (G.P.O). The Telegraph Act of 1868 established the office of the Postmaster General's right to exclusively maintain electric telegraphs. This would subsequently extend to telecommunications and radio broadcasting. The title of 'Postmaster General' was abolished under the Post Office Act of 1969. A new public authority governed by a chairman was established under the name of the 'Post Office.' The position of 'Postmaster General' was replaced with 'Minister of Posts and Telecommunications.'

Radio: currently used a synonym for 'electromagnetic radiation' and also 'wireless.'

Essentially the terms wireless and radio can mean exactly the same thing. Surprisingly it was the word radio that first came into use even before Heinrich Hertz proved the existence of electromagnetic waves, but despite this very early usage it has always been regarded as the modern version of the term wireless.

Communication by electromagnetic radiation became known as wireless simply because there were no wires linking the receiver site to the transmitting station and the term wireless telegraphy was used to separate it from the common cable based telegraph systems sending Morse code messages on a daily basis. The term radio came into being because the communication system was a receiver of radiated electrical signals.

Originally the term 'radio' had been used as a more general prefix meaning 'radiant' or 'radiation'; hence 'radio-activity' for the alpha, beta, and gamma rays that are emitted by decaying atoms. It was based on the verb to radiate (in Latin 'radius' means 'spoke of a wheel, beam of light, ray'.)

In Europe, some of the earliest pioneers investigating Hertz's discovery began to commonly employ the 'radio' prefix to describe the new phenomenon. In 1890 the French physicist Edouard Branly called his coherer receiver device a 'radio-conductor.' He published his findings in 1891, but he made no mention of using his invention to detect electromagnetic waves even though his choice of name for his device was prophetic. On 29th December 1897, *The Electrical Review* reported on 'Hertzian Telegraphy in France' and noted that 'Mr. Branly... calls these receivers ...radio-conducting tubes.'

Other compound usages soon followed, as *The Electrician* for 24th October, 1902 included an article titled 'The Radio-telegraphic Expedition of the H.I.M.S. *Carlo Alberto,*' while 'The Wireless Telegraph Conference', in the 20th November 1903 issue of the same magazine, included numerous references to 'radio-telegrams', 'radiograms', 'radiographic stations' and 'radio-telegraphy', and a report about Belgium's marine applications in the 19th November 1904 *Electrical Review* noted that 'radio-telegraphy has entered into the domain of current practice.'

Eventually, compound terms such as 'radio-telegraphy' and 'radio-telephony' were shortened to just 'radio.' The 1st April 1906 issue of *Telegraph Age* reported that 'the British Post Office... has adopted the word 'radio' as the designation for a wireless telegram', and this practice was adopted internationally later that year by the Berlin Radiotelegraphic Convention, which specified that 'wireless telegrams shall show in the preamble that the service is radio.'

The term 'wireless communication' makes the distinction even more blurred as it does not implicitly imply the use of electromagnetic waves. The earliest pioneers who struggled to communicate without wires used electromagnetic conduction systems, either through water or earth. They included Loomis, Stubblefield, Steinheil, Morse, Lindsay, and Dolbear (even though his system may also have used electromagnetic radiation). Meanwhile Morse and Preece obtained successful results and even went into commercial operation using systems that communicated without wires, but employed electromagnetic induction techniques.

To further confuse things the phrase 'wireless communication' has also been applied to numerous other signalling technologies. Historically these have included smoke signals, drum beats, gun shots, heliographs (flashing mirrors) and flag semaphore. Even today maritime flags and signal lamps are still in use to communicate over short distances and they definitely do not need wires.

It was perhaps in part these issues that led to the disappearance of the word wireless.

Although both terms are correct, the Americans adopted the term radio in common usage earlier than the UK, it was used by American pioneer Lee De forest as early as 1907 and formally adopted by the US Navy in 1912. On 28th September 1923 the BBC officially accepted 'Radio' by publishing the *Radio Times*.

The word 'wireless', when used in relation to broadcasting or radio equipment is now commonly regarded as an old fashioned term. However in recent years the term 'wireless' has gained renewed popularity through the rapid growth of short-range computer networking such as Wireless Local Area Networks (WLAN), Bluetooth and Wi-Fi. Today, the term 'radio' often refers to the actual transceiver device or chip, whereas 'wireless' refers to the system and/or method used for communication, hence one talks about radio transceivers and Radio Frequency Identification (RFID), but about wireless devices and wireless sensor networks.

In this book, I have tended to employ the word wireless, if nothing else, as John Wayne once said, I like the sound of the word. But if sometimes the word radio also appears then that is correct as well. Also I was recently reminded that a young Italian called Guglielmo Marconi once called his Company Marconi's *Wireless Telegraph Company Ltd*, and how could he be wrong?

Receiver: A wireless or radio receiver; the set that receives a wireless message or radio broadcast. Early wireless sets usually consisted of separate receivers and transmitters, while later radio communication sets were transceivers - combination receivers and transmitters.

R.F.: Radio Frequency: a frequency in the range within which radio waves may be transmitted, from about 3kHz to about 3GHz. The frequency spectrum is commonly broken down into:

L.F.: Low Frequency: a radio frequency in the range 30-300kHz. Wavelength (10km-1km) Waveband known as 'Long Wave.'

M.F.: Medium Frequency: a radio frequency in the range 300 kHz – 3MHz.

(Wavelength 1000m-100m.) Waveband know as 'Medium Wave.'

H.F.: High Frequency: a radio frequency in the range 3-30MHz. (3,000 kHz. - 30,000kHz.). Wavelength 100m-10m. Waveband known as 'Short Wave.'

RX: See receiver.

Spark Coil: More commonly referred to as an Induction Coil.

Spark Gap: Electrical apparatus used for the production of repeated spark discharges between the two or more electrodes forming a transmitting device for generating radio frequency (R.F.) waves, commonly used in early wireless transmitter. Remained in common use until the advent of the multiple tuner and magnetic detector combination around 1908. The first demonstrations of practical radio communication, both by Marconi and others, were carried out using spark gap transmitters.

Spark gaps used in early radio transmitters varied in construction, depending on the power to be handled. Some were fairly simple, consisting of one or more fixed (*static*) gaps, usually between brass or metal spheres connected in series, while others were significantly more complex. Because sparks were quite hot and erosive, electrode wear and cooling were constant problems. A spark gap gets very hot quickly and in some cases, the transmitter was fitted with a blower motor to remove the hot gases.

Spark Transmitter: In its simplest form a spark transmitter consists of a spark gap connected across an oscillatory circuit consisting of a capacitor and an inductor in series. The Leyden jar [capacitor] is charged to a high voltage by an induction coil. When the potential across it was sufficiently high to break down the insulation of air in the gap, a spark jumps across the gap.

In normal conditions, the air gap forms a barrier between the two capacitors. To break the gap, a spark must be induced. To this end, the capacitor that is directly connected to a power source is charged with electricity. The electrical charge does not immediately dissipate because of an in-built resistor that 'holds' the charge.

After the electrical charge reaches the minimum threshold voltage, the electric charge passes through the resistor to the conductive electrode touching the air gap. This electrode subsequently releases the charge towards the gap and this ionizes the air within the gap; this forms a connection between the two electrodes at the gap.

This initial spark makes the air gap conductive and the pulse generates an electromagnetic wave in the radio frequency range (around 350,000 Hz). This transforms the spark-gap transmitter into a resonant circuit (an inductor-capacitor setup) where alternating current is conducted. This process is very similar to what happens when the negatively charged particle from the air touches the positively charged particle on land in a lightning flash.

Very soon after the initial spark, though, the wave weakens since some of the energy escapes though the antenna found on the side of the second capacitor and some are used up due to resistance. At a certain point, the ionized gas builds up resistance once again and the flow of current stops. Since the radio wave is weakened after a while, spark-gap transmitters acquired the name 'damped oscillation' transmitters.

Since this spark has a comparatively low resistance (an ohm or two), the spark discharge was equivalent to the closing of an L-C-R circuit. The condenser then discharged through the conducting spark, and the discharge took the form of a damped oscillation, at a frequency determined by the resonant frequency of the spark

By 1905 a 'state of the art' spark gap transmitter operated on 400 metres (750 Khz) and generated a signal from about 250 metres (1.2 Mhz) to 550 metres (545 Khz).

Although these wireless stations were terribly inefficient compared to modern standards, these transmitters were able to reach distances out to 100 miles (160km) using a 15inch (38cm) spark coil and a kilowatt station. Professional installations like ships at sea used transmitters up to 5 kW and reached distances up to 500 miles (800 km).

There was one other notable characteristic about the spark gap transmitters at this time. On reception, each signal sounded just a little bit different from the rest as every spark station had its own characteristic sound.

This signal characteristic was usually determined by the electrode gap spacing, electrode shapes, operating voltage and power levels inherent to each hand made transmitter. With a little practice, an operator could identify a transmitting station based on its characteristic sound in his headphones. Without the ability to tune wireless equipment successful operators had to learn to distinguish wanted signals from unwanted ones by using the 'cocktail party effect' whereby one can listen to one conversation, ignoring other simultaneous and maybe even louder ones.

However from an operational security viewpoint this was not good for any navy, as a ship could easily be identified simply by the tone or sound of its transmitted signal.

By 1912 spark transmitters were being replaced by other technologies but their use, despite their widespread interference, was not officially banned until 1st January 1930.

Syntony: The condition of being syntonic or 'tuned' so as to respond to one another, as two electric circuits. It was used as an early term to describe 'tuned' circuits, as developed by Braun, Lodge and Marconi around 1900.

Tapper: An electromagnetic device used to shake the filings of a coherer after the detection of a wireless signal has caused the filings to *cohere* and become conducting. The tapper essentially reset the coherer after each signal reception so it was able to detect the next incoming wave. Sometimes referred to as a Morse Tapper. See also Coherer.

Telegraphy: The operation of telegraph apparatus, a system for sending messages (originally by Morse code) through a wire, or by means of radio waves.

Telephony: In the early days, the transmission of sound by radio was regarded simply as a new means of sending messages. Its original name, Wireless Telephony, indicates that the system was widely considered to be just a new kind of telephone. The term is now used for the science of communication by any device able to transmit intelligible speech over distances.

Transmitter: A wireless or radio transmitter; the set that transmits or sends a wireless message or radio broadcast. Early wireless sets usually consisted of separate receivers and transmitters, while later radio communication sets were transceivers - combination receivers and transmitters.

TX: See Transmitter.

Wavelength: The distance between successive peaks or maxima of an electromagnetic wave. The wavelength determines the nature of the various forms of radiant energy that comprise the electromagnetic spectrum.

Wavemeter: An electrical device for measuring the wavelength of a radio frequency (R.F.) wave either directly or indirectly, through the determination of the frequency.

Wireless: See Radio

W/T: Wireless Telegraphy.

Wireless Telegraphy: An historic term used today as applied to early radio telegraph communications techniques and practices. Wireless Telegraphy originated as a term to describe electrical signalling without the use of electric wires to connect the end points. The intent was to distinguish it from the conventional electric telegraph signalling of the day that required wire connection between the end points.

Wireless telegraphy rapidly came to be synonymous with Morse code transmitted with electromagnetic waves decades before it came to be associated with the term radio. Wireless telegraphy is used widely today by amateur radio hobbyists where it is commonly referred to as continuous wave (CW) radio telegraphy, or just CW.

Wireless Telephony: The transmission of speech rather than Morse code.

Bibliography
and
Further Reading

- Baker, W.J., *A History of the Marconi Company*, Methuen & Co., London (1970).

- Vyvyan, R.N., *Wireless Over Thirty Years*, George Routledge & Sons, London (1933).   Reprinted as *Marconi And Wireless*, EP Publishing Limited, Yorkshire, England (1974).

- Kemp, George, '*Extracts from the diary of G. S. Kemp.*' *Marconi Archives*, Marconi plc.

- Maria Christina Marconi, '*Marconi My Beloved*' ISBN 0-037832-39-1 Dante University of America Press - 2001

- *Watchers of the Waves by* Brian Faulkner. G.C. Arnold Partners. 1996. ISBN 1898805 09 1

- Jacot, B. L., and D. M. B. Collier. '*Marconi: Master of Space. London*': Hutchison, 1935.

INDEX

Figures in *italics* refer to illustrations.

Hertz, Heinrich   xxiii, 41ff
Hurst Castle   171, 173ff

Induction Coil   6, 8, 15, 39, 41, *42*, 58, 66, 82, 66, 82, 90, 115, 121, 178, 190ff
Induction System, Preece's   xxiii, 26, 172ff

Jackson, Henry B.   17, 23, 63, 180
Jameson-Davis, Henry   13, 14, 26ff, 34, 62, 91, 105, 106
Jigger   88, 89, 96, 110, 111, 114, *115*, 120, 121, 192ff

Kelvin, Lord   58ff, *60,* 63
Kemp, George   16, 17, 18, 19, 23, 25ff, 37, 38, 42, 43, 44ff, 50ff, 65, 68ff,
75, 81, 92, 97, 98ff, 119, 124, 138, 141, 177
Kennedy R.E, Captain J.N.C.   46, 50

Lloyds Signalling Station, Niton   152ff
Luttrell's Tower   167ff, *170*

Madeira House wireless station   51, 64ff, 81, 92, 156ff
Maxwell, James Clerk   xxiii, 4, 5, 188

Navy, British   17, 18, 22, 23, 26, 30, 50, 63, 73, 97, 102, 108, 117, 162, 165,177
Needles Hotel, Alum Bay   26ff, 131ff
Niton wireless station   109ff, 146ff
Nobel Prize   36, 116

Obach Cell   39, 67, 82, 192
Osborne House   vii, 73ff, 97, 132, 145, 167, 171
Osborne (Royal Yacht)   73ff, 75

Post Office   13, 15, 16, 20, 22, 23, 25, 30, 50, 52, 59, 63, 91, 97, 153, 172, 187
Preece, William   13, *14,* 15, 16, 20, 21, 23, 24, 25, 26, 30, 50, 59, 63, 98, 99, 122,
171, *175*, 194
Prince of Wales   74ff

Queen Victoria   vii, 12, 73ff, 76, 78, 97, 111, 145, 167, 171ff, *175*

Royal Needles Hotel – see Needles Hotel
Royal Sandrock Hotel – see Sandrock Hotel

The author photographed in front of Isle of Wight Niton station.
Seated on the aerial base, May 2011

About the Author – Tim Wander

Raised and educated in Melton Mowbray in Leicestershire, an Honours Degree in Computer Science from Aston University in Birmingham brought him by chance to work at GEC-Marconi Communication Systems' Writtle site, near Chelmsford in Essex. He spent the first 17 years of his career with various arms of the GEC-Marconi Company worldwide, designing, developing and managing radio, telecommunication and control system projects. Tim left Marconi's in 1999 and spent three more years in senior management within the electronics industry in the City of London. Tim then decided on a career change, finding time to restore a number of classic Jaguar E type cars, he then became a Project Manager for a series of major building projects around the world.

Tim has written numerous other books, including '**Marconi on the Isle of Wight'**, produced in 2000 for the centenary of the closure of the Alum Bay, Royal Needles Hotel station. A long held interest in early radio sets inherited from his father and a passion for the early days of radio broadcasting had led him to write **'2MT Writtle - The Birth of British Broadcasting'**, first published in 1988. After 22 years the second, completely rewritten and much larger definitive edition was published in October 2010. In 2012 he published the story of the world's first purpose built wireless factory – **'Marconi's New Street Works 1912-2012. Birthplace of the Wireless Age**.'

He has also written several radio plays based on the New Street and Writtle broadcasts and a TV script is in production. More books are planned and you can keep track of Tim's new and past titles and contact the author via *2MTwrittle.com*. Tim's hobbies still include a passion for early Jaguar cars, including a period of historic motor racing. Tim is now a Freelance Author, Lecturer, Consultant and Project Manager. Married to Judith with three children, Michael, David and Elizabeth, Tim spends much time on the Isle of Wight. Indeed Judith's family hail from the Island.

Lightning Source UK Ltd.
Milton Keynes UK
UKOW04f0103070417

298565UK00009B/162/P

9 780755 207206